D0277101

The Comedy of Error

Jonathan Silvertown is Professor of Evolutionary Ecology in the Institute of Evolutionary Biology at the University of Edinburgh. He is the author of seven previous books.

The Comedy of Error

why evolution made us laugh

Jonathan Silvertown

SCRIBE
Melbourne • London

Scribe Publications
2 John St, Clerkenwell, London, WC1N 2ES, United Kingdom
18–20 Edward St, Brunswick, Victoria 3056, Australia

First published by Scribe 2020

Copyright © Jonathan Silvertown 2020

All rights reserved. Without limiting the rights under copyright reserved above, no part of this publication may be reproduced, stored in or introduced into a retrieval system, or transmitted, in any form or by any means (electronic, mechanical, photocopying, recording or otherwise) without the prior written permission of the publishers of this book.

The moral right of the author has been asserted.

Every effort has been made to acknowledge and contact the copyright holders for permission to reproduce material contained in this book. Any copyright holders who have been inadvertently omitted from the acknowledgements and credits should contact the publisher so that omissions may be rectified in subsequent editions.

Typeset in Adobe Caslon Pro by the publishers

Printed and bound in the UK by CPI Group (UK) Ltd, Croydon CR0 4YY

Scribe Publications is committed to the sustainable use of natural resources and the use of paper products made responsibly from those resources.

9781913348182 (UK edition)
9781922310095 (Australian edition)
9781925938456 (ebook)

Catalogue records for this book are available from the National Library of Australia and the British Library.

scribepublications.co.uk
scribepublications.com.au

For Rob, a most dedicated friend

LONDON BOROUGH OF RICHMOND UPON THAMES	
90710 000 452 906	
Askews & Holts	27-Nov-2020
152.43 SIL	
RTH	

DISCARDED FROM RICHMOND UPON THAMES LIBRARY SERVICE

Contents

Chapter One

Comedy and Error

'It's a delightful thing to think of perfection, but it's vastly more amusing to talk of errors and absurdities.'
Fanny Burney (1752–1840)

There is comedy in errors. Shakespeare showed us so, although the connection between error and humour had been recognised for millennia. The Bard took the plot for his *Comedy of Errors* from the Roman playwright Plautus, amplifying the farcical effect of the original by adding a second helping of mistaken identity between his characters.[1] But 21st-century science has discovered something genuinely new about the comedy of errors that neither Plautus nor

Shakespeare could ever have conceived.

It turns out that errors are much more than just a plot device for humorous tales — they are the very essence of what we find funny. There is an area in the human brain that is specifically dedicated to detecting errors. These errors are processed, compared with expectation, and those judged humorous ricochet around the brain, producing laughter. Suddenly, with this discovery, the two cultures of science and art have collided and, like strangers meeting in a pub, we find them bonding over jokes. This book is witness to that unexpected and fruitful encounter. We'll probe the questions it throws up, catch the jokes that fly out, and find deeper meaning in frivolity.

Why are some errors funny and others not? Why is laughter involuntary and infectious? Laughter is found in all cultures and when heard it is recognisable across boundaries of language. Babies laugh and neither eyesight nor hearing is required to acquire the behaviour.[2] All these characteristics strongly suggest that laughter is hard-wired into the human psyche, and to an evolutionary biologist like me that immediately provokes my favourite question: what good is it? Answering that question is the ultimate purpose of this book. Why did evolution make us laugh?

Though an evolutionary biologist, I tiptoed into this territory as an interloper, more used to interrogating the whys and wherefores of plants than of minds. What I discovered was that from Aristotle (384–322 BCE)[3] onwards, (almost) anybody who wants to be taken seriously has written about laughter: Henri Bergson, Charles Darwin, René Descartes, Sigmund Freud, Thomas Hobbes, Immanuel Kant, Artur Schopenhauer, to name only the most hilarious. 'There are few things less entertaining than academics pontificating about laughter,'[4] as one more recent writer said, before proceeding to prove her point by doing just that. Pontificating can pay though:

How does the Pope pay his bills?
PayPal.

What have the professors of fun been up to? Do they wear clown shoes and baggy pants? And if they do, how can you tell them from ordinary professors? But I digress. In the *Primer of Humor Research*, the editor and godfather of humour scholarship calls trespassers like me 'first-timer pests',[5] and abhors our weakness for jokes. It's a strange world in which the scholars are afraid that you will laugh, while the performers are scared that you won't. In the academic reference *Handbook of Humor Research*, the editors lament that: 'For reasons that remain unclear, many investigators published only one or two humor studies before abandoning the area in favour of some other research domain.'[6] Perhaps they were scared off by the godfather's scathing pen? I've noticed the same lack of fortitude among scientists who study slugs. Laughter and molluscs seem equally fatal to an academic career. Some subjects, it seems, are better not taken too seriously.

> *A man walks into a cinema, sits down, and notices that there is a large slug sitting in the seat next to him.*
>
> *'What are you doing here?' asks the man in surprise.*
>
> *'Well, I loved the book,' replies the slug.*

Which goes to prove that neither slugs nor jokes about them get us anywhere. An awful lot of blind alleys have been explored on the long road to understanding humour.

Back in the lab, a paper on how to get robots to be funny begins, 'First, laughter has a strong connection with humour.'[7] 'No shit, Sherlock!', you might say, but there is a serious distinction to be made. We should distinguish between humour — the stimulus, and laughter — the response. These are separate things and either may occur without the other, as any stand-up comedian knows only too well. Sir Ken Dodd (1927–2018) defined the craft of comedy, of which he was a consummate master, as 'the performance of humour to obtain laughter'.[8] There will be jokes, perhaps about slugs, that you recognise as humorous, but that don't make you laugh out loud. Conversely, a tickle can elicit laughter without the stimulus of humour. What tickles your fancy can be quite revealing.

What's the difference between erotic and kinky?
Erotic is using a feather. Kinky is using the whole chicken.

We all instinctively know two things about laughter: one, it is a social phenomenon and two, it is not just about humour, but happens when humans are having fun. Psychologist Robert Provine listened in on people's conversations and discovered that most laughter occurs in ordinary discourse and not, as one might have imagined, just when someone says something funny.[9] You can test this out for yourself in any bar or social situation that lends itself to unobtrusive ear-wigging. I've found it to be true. Charles Darwin knew it. Writing in 1872, he commented that when young people past childhood 'are in high spirits, there is always much meaningless laughter'.[10]

Can we work out how humour works and why we laugh at it? Should we even try, or is analysing a joke like using a pin to explain how a balloon works? Why does explanation deflate rather than enhance a joke? There is a scientific explanation that we shall explore later. However, there is also a romantic notion that the moment we try to analyse a thing of beauty or joy, we destroy it, much like a vivisectionist investigating a throbbing heart with a scalpel. I write in the conviction that the very opposite is true — understanding increases rather than diminishes pleasure. This book is a test of that proposition.

Although most laughter happens spontaneously and not in response to humour, jokes are my scalpels in this book. They are selected to make you first laugh and then think. There is actually a prize, called the Ig Nobel, for scientific research that does the same thing. In 2018, the Ig Nobel Prize was won by a team of surgeons in Portland, Oregon for 'using postage stamps to test whether the male sexual organ is functioning properly'. It makes you wonder what these guys think the proper function of the male sexual organ is. Actually, they used stamps to devise an inexpensive method for diagnosing erectile dysfunction during sleep.[11] Well, only inexpensive if you use second class stamps, of course. You create a collar of stamps that fits snugly around said organ before you go to bed. If you wake up in the morning with the collar torn along the perforations, you can turn over and wake your partner with the good news.

Who said philately will get you nowhere?

Here is the plan of the book. It's cunningly simple. You know what Chapter One is about: you just read it. In Chapter Two we will hunt out humour, wrestle it to the ground and pin it down with a definition. This has been tried before and humour always escapes like a will-o'-the-wisp, but I am going to sneak up on it while it is looking the other way and tickle it into submission. In Chapter Three, with humour wriggling under the microscope, we take a good look at our prey and see what it is made of. Chapter Four gets to the bottom of the tickle, the original way that evolution made us laugh, and we find out why laughter is contagious. Chapter Five gets face-to-face with smiling and in Chapter Six we find out what good laughter is to evolution. Finally, with the messy biology of laughter cleaned up, in Chapter Seven we shall see what culture has built upon the biological foundations. We defiantly scrape the bottom of the barrel of fun and find out what makes Deaf jokes, musical jokes, and Jewish jokes different and funny. One humour researcher has written disapprovingly that 'because the subject is humor, many people see the field as an opportunity to tell jokes'.[12] Imagine that! You have been warned.

Chapter Two

Humour [illegible]

[illegible]

Chapter Two

Humour and Mind

The quest for the essence of humour is older than the alchemists' search for the philosopher's stone. Even great comics like W.C. Fields (1880–1946) haven't been able to figure it out:

> The funniest thing about comedy is that you never know why people laugh. I know what makes them laugh but trying to get your hands on the why of it is like trying to pick an eel out of a tub of water.[1]

The result is that there are dozens and dozens of theories about what makes humour funny — at least 100 and counting.[2] Like the famous Buddhist parable about six blind men trying to understand what an elephant is like, most humour theories only grasp a part of the truth about the beast. In the parable, the first blind man collided with the elephant's side and pronounced the animal like a wall. The second felt a tusk and decided an elephant is like a spear; the third found the coiling trunk and thought the elephant was like a snake; the fourth touched a sturdy leg and said it was a tree; the fifth was swatted by an ear and believed the elephant to be like a fan, and the sixth seized the swinging tail and concluded an elephant is very like a rope. In the end, according to a rhyme version of the parable by American poet John Godfrey Saxe:[3]

And so these men of Indostan
Disputed loud and long,
Each in his own opinion
Exceeding stiff and strong,
Though each was partly in the right,
And all were in the wrong!

To fathom the nature of the comedy beast we need to observe it in its natural habitat, so join me in the city of Edinburgh. During the Fringe Festival each August, every conceivable style and flavour of comedy is to be found here, alongside music, drama, art, and literature. If an essence of humour exists in all the diversity of comic expression, this is going to be where we will find it.

The Pleasance, one of the biggest laughter venues, is right on the edge of Edinburgh's Holyrood Park. We press through the crowds in the Pleasance courtyard, thronged with drinkers and people enjoying Scottish delicacies like vegan haggis, Cullen skink, and Aberdeen Angus beef burgers. The aromas waft and linger in the damp evening air of a Caledonian August. In the 18th century, the Pleasance was the haunt of prostitutes on the edge of town. From edge then to Fringe now, this is still the place for cheap thrills, but today they are to be found in more than two dozen theatre spaces running shows simultaneously, every hour, from morning to night.

The choice in this mother lode of laughs is overwhelming. On a poster wall displaying what's on offer, it looks like every one of the shows advertised is on a list of unmissable hilarity. A mime show called *Fishbowl* by a French troupe evokes laughter about the human condition, despite touching on issues of loneliness and friendship, all in total silence. Matt Winning, a climate scientist and stand-up comic, manages to wring laughs from the climate emergency and still leave his audience with the clear message that the evidence demands action by us all now. Other performers share the humour they find in their outsider status as migrants, feminists, people of colour or LGBTQ+.

To my disappointment, political satire is not much in evidence this year, perhaps because today's prominent politicians, most notably Donald Trump, satirise themselves, and anyone who wants one has a front-of-house seat via Twitter. One of the few satirical comics performing this year prefers to target her audience. Her blurb reads: 'Titania McGrath is a radical intersectionalist poet committed to feminism, social justice and armed peaceful protest. As a millennial icon on the forefront of online activism, Titania is uniquely placed to explain to you why you are wrong about everything and how to become truly woke.'[4] The plaudits for the show come from the right-wing press.

The diversity of performances at the Pleasance suggests that there is no limit to what people find funny. Can that be true? Is the essence of humour a figment of the imagination, like the philosopher's stone? When the Greek philosopher Aristotle approached the elephantine body of humour, he grasped it by the butt and concluded that its essence is ridicule. 'Comedy aims at representing men as worse, Tragedy as better than in actual life,'[5] he wrote, perhaps with conscious irony. Tragically, a whole book that Aristotle wrote about comedy is lost. This is possibly why the English comic Arthur Smith observed, *'There are three basic rules for great comedy. Unfortunately, no one can remember what they are.'*

However, Aristotle must certainly have been familiar with the joke that the hero Odysseus plays upon Polyphemus in Homer's *Odyssey*. Odysseus and his men take refuge in a cave on the island of Sicily, not realising that the cave belongs to a man-eating giant with one eye — the cyclops Polyphemus. The cyclops returns to his cave and eats six of Odysseus's men, but Odysseus offers him wine and gets the giant drunk. The intoxicated Polyphemus promises Odysseus a gift if he will tell him his name. Odysseus replies that his name is 'Nobody' and the giant says that in reward he will eat Nobody last. Polyphemus falls into a drunken stupor and while he is asleep, Odysseus drives a wooden stake into the cyclops' eye. Polyphemus yells out for help to his fellow giants, crying 'Help! Nobody has blinded me!' Hearing this and thinking that Polyphemus must be having a bad dream, the other giants of course ignore him. The next morning, Odysseus and his surviving men escape from the cave by clinging beneath the bellies of the blind cyclops' sheep as he ushers them out to graze.

The oldest surviving joke book, called *Philogelos*, which is Greek for 'The Joker', gives us some idea of what made the ancient Greeks laugh.[6] The surviving version of *Philogelos* was written in Latin around the time that the Romans retreated from Britain. Maybe the Romans desperately needed cheering up as they fled the rain for the sun-drenched shores of Italy? Perhaps. Though written in Latin, the jokes, like so much Roman culture, have Greecy fingerprints all over them. *Philogelos* is full of xenophobic jokes about the stupidity of people whom the Greeks despised, like inhabitants of the city of Sidon in Phoenicia:

> *A Sidonian lawyer is chatting with two friends.*
>
> *One says, 'It's not right to slaughter sheep because they give us milk and wool.'*
>
> *The other says, 'It's not right to kill cows because they give us milk and pull our ploughs.'*
>
> *Then the lawyer adds, 'And it's not right to kill pigs either because they give us liver, bacon, and pork chops.'*

Other butts of the jokes in Philogelos are curmudgeons, cowards, fat people, simpletons, apprentices, and slaves:

> *I bought a slave and he died. I went back to the slave dealer and said, 'I bought a slave from you and he died.' Know what the dealer said? 'Don't blame me. He never did that when I had him.'*

Humour has come a long way since ancient Greece. Or has it? Isn't the dead slave joke just an ancestor of

Monty Python's famous dead parrot sketch? As one gag writer has said, there are no old jokes, just the ones you have heard before.

Laughing at others is a feature of a great deal of humour. To the English philosopher Thomas Hobbes (1588–1679),[7] laughter was a 'sudden glory' in the realisation that one is superior to the person who is the butt of a joke. Odysseus certainly felt sudden glory over the cyclops as he taunted the blind giant by yelling out his real name as he escaped to sea in his boat. To this day, you can see the rocks off the coast of Sicily that, according to fable, Polyphemus flung into the sea after Odysseus's retreating ship.

Slapstick, the mainstay of Laurel and Hardy films, Bugs Bunny and the rest, is an innocent variant of Hobbesian humour. Watching Stan and Ollie trying to manoeuvre a piano up a steep staircase is hilarious, but only if you believe they are not in real danger of being crushed by its weight. Similarly, in their cartoons, Tom and Jerry are perpetually at war with each other, but if the cat ever actually caught and ate the mouse there would have been no laughter. The term 'slapstick' comes from an actual stick called a *bataccio* that was used in 16th-century Italian *commedia dell'arte*. It was made of two pieces of wood hinged together so that they make a loud slapping sound when used to comically beat a stooge. It's important for the comic effect that there is no real injury.

Even if actual physical harm is not funny to most people, the humour of superiority does have victims. Sexist and racist jokes are two inglorious genres of Hobbesian humour that typically invite the listener to laugh at the ugliness of mothers-in-law, or the stupidity of the Irish, or blondes. You've heard the jokes and I am not going to illustrate them here. Happily, feminist jokes are turning the tables on male chauvinism. The difference between male and female perspectives can turn on subtleties of grammar:

An English teacher writes the sentence 'A woman without her man is nothing' on the blackboard and asks his students to punctuate it. A male student writes, 'A woman, without her man, is nothing.' A female student writes, 'A woman: without her, man is nothing.'

Or, if mathematics is more your style, try this one:

Two mathematicians were having dinner in a restaurant, arguing about the average mathematical knowledge of the public. One claimed that this average was woefully inadequate, whereas the other maintained that it was surprisingly high. 'I'll tell you what,' said the cynic, 'ask that waitress a simple maths question. If she gets it right, I'll pay for dinner. If not, you do.' He then excused himself to go to the bathroom, and the other called the waitress over. 'When my friend comes back,' he told her, 'I'm going to ask you a question, and I want you to respond "one-third x cubed". There's twenty pounds in it for you.' She agreed. The cynic returned from the toilet and called the waitress over. 'The food was wonderful, thank you,' he said, and the other mathematician started, 'Incidentally, do you know what the integral of x squared is?' The waitress looked pensive; almost pained. She looked round the room, at her feet, made gurgling noises, and finally said, 'Um, one-third x cubed?' So the cynic paid the bill. The waitress wheeled around, walked a few paces away, looked back at the two men, and muttered under her breath, '... plus a constant.'[8]

Every joke involves a punchline that upsets a preconceived notion. The joke itself may establish the preconception with a set-up such as the story in the last joke about the wager between two mathematicians. In a one-liner, the set-up may already be in the culture and the punchline exploits that. In what might be the best put-down ever, the writer Rebecca West called a chauvinist fellow writer not '*every inch a gentleman*' but '*every other inch a gentleman*'—inserting just one word to turn a cliched compliment into a witty feminist barb against his manhood.[9]

Jokes that work on a set-up that you are already carrying in your head can subversively expose prejudice in the hearer:

What do you call a black aircraft pilot?
A pilot.

The classic comic movie *Blazing Saddles* by Mel Brooks is essentially built around the same joke, 'What do you call a black sheriff: Sheriff.' The result is an hilarious satire on the genre of the Western.

Satire, from Jonathan Swift's *Gulliver's Travels* mocking the politics and society of 18th-century England to *The Daily Show*, which does the same in 21st-century America, is an entire genre of Hobbesian superiority humour that tens of millions glory in every week. Satirical humour differs from the jokes about minorities and women because it punches up, not down. David Levi was an Israeli politician in the 1980s who was mercilessly taunted by a whole cycle of jokes about his stupidity. The recipe works equally well with local substitutes, of course.

One day, Levi's secretary hears on the news that there is a traffic hazard on a route that she knows her boss is using to get home. She calls him up on his mobile and says, 'David, there's a news report that there is a maniac driving the wrong way on the motorway. You'd better watch out!'

'Tell me about it,' says David Levi. 'There's not just one of them, there's hundreds of the maniacs!'

Eventually, the joke cycle finds its way to Levi himself:

A man turns to a fellow passenger on a plane and thinks he'll open the conversation with a joke.

'Hey, have you heard the latest one about David Levi?' he asks.

'Huh?' says his companion. 'What do you mean? I am David Levi.'

'That's all right,' says the man. 'I'll speak slowly.'

Hobbes observed that '[W]hatsoever it be that moveth laughter, it must be new and unexpected.' Novelty and surprise are important to humour and we shall see why later, but significantly a joke need only be new to the hearer — it does not have to be new in the sense of 'contemporary'. Even if you have no idea who the 19th-century British politicians Benjamin Disraeli (1804–1881) and William Gladstone (1809–98) were, what the former said about the latter is still funny *if* you have not heard it before:

'The difference between a misfortune and a calamity is this: if Gladstone fell into the Thames, it would be a misfortune, and if someone hauled him out again, that would be a calamity.'[10]

There are some perennial objects of scorn like lawyers, where the butt of the Hobbesian joke is familiar, but their torment may not be:

Four surgeons were taking a coffee break and were discussing their work. The first said, 'I think accountants are the easiest to operate on. You open them up and everything inside is numbered.'

The second said, 'I think librarians are the easiest to operate on. You open them up and everything inside is in alphabetical order.'

The third said, 'I like to operate on electricians. You open them up and everything inside is colour-coded.'

The fourth one said, 'I like to operate on lawyers. They're heartless, spineless, gutless, and their heads and their arses are interchangeable.'[11]

Hobbesian jokes can stay ever fresh — sealed within a time capsule in which two protagonists vie with each other, one eventually besting the other with an ageless put-down.

A man went to his barber for a haircut and of course, as it always does, the subject of the conversation turned to up-coming vacations.

'We are going to Italy for a week,' said the customer, 'and my wife and I are really looking forward to it.'

'Italy?' said the barber. 'You don't want to go there! It's horribly hot and crowded and the food is awful.'

'It's too late to change our plans now and anyway my wife is dying to see Rome.'

'Rome?' said the barber. 'You really don't want to go there. The traffic is terrible and the whole place is falling down! Where are you staying, anyway?'

'We've got an executive suite booked at the Rome Hilton,' said the man.

'Don't stay at the Hilton!' said the barber. 'I stayed there once and it was the worst experience of my entire life.'

A month goes by and the customer is back in the barber's chair, tanned from his holiday in Italy. 'How did it go?' asks the barber. 'I bet it was awful, wasn't it?'

'Well, we went to Rome and we visited the Vatican. While we were there, a priest beckoned us into the Sistine Chapel.'

'The Sistine Chapel?' says the barber. 'Overrated.'

'And then,' says the man, 'a small door opened, and the Pope came out and spoke to us.'

'Really? What did he say?'

'"My son,"' he said to me, "every Sunday I stand on the balcony overlooking St Peter's Square and I see thousands of heads beneath me as I give the crowd my blessing." And, then the Pope said to me, "In all the Sundays, in all the many years that I have been Pope, I have never, ever seen a haircut as bad as yours."

Sometimes, Hobbes observed, one laughs at a previous version of oneself to demonstrate that one is now superior to some earlier foolishness. Here is a joke along those lines that my father once brought home after a visit to the barber:

A cannibal went on vacation for a fortnight. His friends saw him off as he paddled his canoe away downstream and waved until he was lost to sight. Two weeks later, the friends, eager to hear how it went, were at the riverbank as he reappeared around the bend. He beached his canoe, stood on one leg, and with the help of a branch used as a crutch, the cannibal hopped ashore. His friends were horrified to see that their friend now only had one leg. 'What happened?!' they cried. 'Oh, it was a marvellous holiday,' replied the cannibal, 'but it was self-catering.'

It's just a short hop from a Hobbesian joke like this to the humour of self-deprecation. The American comedian Rodney Dangerfield (1921–2004) built a very successful career in stage and TV from jokes made at his own expense. His catchphrase was 'I don't get no respect.'

I'm a bad lover. I once caught a peeping Tom booing me.

I'm bisexual. I have sex twice a year.

A girl phoned me the other day and said, 'Come on over. There's nobody home.' I went over. Nobody was home.

Last night my wife met me at the front door. She was wearing a sexy negligee. The only trouble was, she was coming home.

Dangerfield would not have been surprised to learn that, according to one study, self-deprecating humour is used by dumber, less successful comedians.[12] It is hard to get self-deprecating humour to work well because it relies on the construction of a lugubrious comic persona. These are not jokes you would tell your mates in the pub, or not if you don't want to buy the next round, anyway. Scottish comedian Arnold Brown has perfected this style:

I enjoy using the comedy technique of self-deprecation — but I'm not very good at it.

Here's an entry-level self-deprecation joke suitable for the comedian on the bottom rung of the professional ladder:

I have a stepladder. It's a very nice stepladder, but it's sad that I never knew my real ladder.

The best joke of this type has to be this one from the English entertainer Bob Monkhouse (1928–2003):

They all laughed when I said I wanted to be a comedian. They're not laughing now.

There is no doubt that Aristotle and Hobbes nailed an important feature of humour with the superiority theory, but between them they only got hold of a foot and a butt of the whole beast. A complete theory of humour needs to explain why puns that play on words are funny, despite lacking any victim or the butt of the joke. Indeed, people may be totally absent, as in these examples:

Time flies like an arrow, fruit flies like a banana.

If a pig loses its voice, is it disgruntled?

What's the difference between unlawful and illegal? Unlawful means against the law, illegal is a sick bird.

I went to the zoo, and when I got there all they had was a dog. It was a shih tzu.

Puns can be bilingual:

An English cat called One-two-three had a swimming race with a French cat called Un-deux-trois. Which cat won? The English cat, because Un-deux-trois cat sank.[13]

According to Freud, what comes between fear and sex? Fünf.

If superiority is not essential to humour, is there another essential ingredient in all the things that make us laugh? Sigmund Freud (1856–1939), inventor of psychoanalysis, was a great collector of jokes. He believed that jokes licence us to liberate hidden psychological meanings that would normally be taboo.

A psychoanalyst asks his patient how his visit to his mother went.

The patient says, 'It didn't go at all well. I made a terrible Freudian slip.'

'Really,' says the analyst, 'what did you say?'

'What I meant to say was "Please pass the salt." But what came out was, "You bitch, you ruined my life!"'

Hence the definition of a Freudian slip as *'When you say one thing but mean your mother.'* But when it comes to the unconscious, there is always the problem of objective interpretation:

A psychoanalyst shows a patient an inkblot and asks him what he sees. The patient says: 'A man and a woman having sex.'

The psychoanalyst shows him a second inkblot and the patient says, 'It's another man and woman having sex.'

The psychoanalyst says, 'You are obsessed with sex!'

The patient replies, 'What do you mean, I'm obsessed? You're the one with all the dirty pictures.'[14]

Sometimes, Freud said, a cigar is just a cigar (Groucho Marx, another cigar aficionado, would have agreed). Freud wrote a book on *Jokes and their Relation to the Unconscious* in which he aimed to distil a single, psychoanalytical theory from observations made by others in their disparate attempts to define the essence of humour. Here's a frequently recycled joke that Sigmund Freud collected:

A man at the dinner table dipped his hands in the mayonnaise and then ran them through his hair. When his neighbour looked astonished, the man apologised: 'Oh! I'm so sorry. I thought it was spinach.'[15]

Freud wanted his psychoanalytic theory to account for the way that jokes playfully juxtapose sense and nonsense, first puzzling and then enlightening the listener. These are indeed elements that can be recognised in most jokes. The set-up is the puzzle and the punchline is the enlightenment. Freud concluded that the pleasure in jokes comes from a kind of mental catharsis delivered by the punchline. He laboriously dissected a squadron of jokes, for example pulling the logic in this one apart:

A gentleman entered a pastry-cook's shop and ordered

a cake, but he soon brought it back and asked for a glass of liqueur instead. He drank it and began to leave without having paid. The proprietor detained him.

'What do you want?' asked the customer.

'You've not paid for the liqueur.'

'But I gave you the cake in exchange for it.'

'You didn't pay for that either.'

'But I hadn't eaten it.'

Freud wasn't sure if this exchange between customer and shopkeeper was really a joke at all, which suggests that his preconceptions about jokes may have got the better of his sense of humour. But then, to be fair, he was writing before the Marx brothers turned frenetic repartee into a successful family business:

One morning I shot an elephant in my pyjamas.
How he got in my pyjamas, I don't know.[16]

Groucho Marx saw what Sigmund Freud had missed — that 'Humour is reason gone mad.' Nothing is more madcap than the best of the Marx Brothers' movies. Though humour can undoubtedly be cathartic, according to Freud's theory people would laugh less and less during a Marx Brothers' movie as their sense of humour is purged. In reality, the reverse happens. Every comedian recognises the need to warm up an audience. Once you are in the mood, you are much more likely to laugh at the next gag. To raise audiences to heights of mirth, comedians aim to keep the gags coming thick and fast. Ken Dodd's yardstick was six gags a minute and he could keep this going for what seemed like for ever. When he saw a member of the audience checking his watch, he said, 'You don't need a wristwatch for my shows, you need a calendar.' The Guinness World record holder for the number of jokes in an hour is Tim Vine, who managed to get a laugh from his audience 499 times in 60 minutes.[17] Vine specialises in silly puns, often with an ironic twist:

Conjunctivitis.com — That's a site for sore eyes.[18]

Someone actually complimented me on my driving today. They left a little note on the windscreen. It said, 'Parking Fine.' That was nice.

So I was reading the obituary column. It said, 'Mars bar, packet of Rolos, Double Decker.' Then I realised that I was reading the 'a bit chewy' column.

Freud's theory of humour is proved wrong by peoples' unquenchable ability to laugh and laugh and laugh. The Marx Brothers' humour, on the other hand, illuminates with a brilliant beam of limelight a property that does seem universal: incongruity. The set-up points in one direction, the punchline points in another and the incongruity between the two is resolved by mad reasoning. You will find incongruity of some kind in every joke in this book and, I would bet, every joke you can think of. With his characteristic insight, Charles Darwin put his finger on it 150 years ago when he wrote that, despite the complexity underlying why adults laugh:

> [S]omething incongruous or unaccountable, exciting surprise and some sense of superiority in the laugher, who must be in a happy frame of mind, seems to be the commonest cause. The circumstances must not be of a momentous nature.[19]

Aristotle possibly understood the importance of incongruity to humour,[20] but the philosopher Immanuel Kant (1724–1804) is usually credited with introducing the idea in the *Critique of Judgement*, published in 1790.[21]

Kant. Now there's a name not to conjure with, or at least not out loud. Sidney Morgenbesser, a professor of philosophy at Columbia University in New York, was once lighting up his pipe as he left the subway when he was pulled up by a policeman. Morgenbesser protested that although smoking was banned in the station, he was actually outside. 'OK, OK,' said the cop, 'but if I let you get away with it, I'd have to let everyone get away with it.' To which Morgenbesser replied, 'Who do you think you are — Kant?'[22] Hours later, Morgenbesser had to be rescued from the local lock-up by another professor of philosophy who explained to the police that the four-letter word used by the prisoner was a proper noun, not an expletive.

Perhaps it was unreasonable of Morgenbesser to expect the policeman at the subway to recognise his reference to Kant's Categorical Imperative which implies that all persons should be treated equally under the law. Not everyone is a big fan of Kant. The philosopher Bertrand Russell is alleged to have observed:

> *Philosophers before Kant had a tremendous advantage over philosophers after Kant in that they didn't have to waste time studying Kant.*

Kant's theory of humour holds that laughter occurs when 'a tense expectation is transformed into nothing'. It's unlikely that the policeman who had to transform Morgenbesser's punishment into nothing ever got the joke, since explanations are rarely funny. Kant illustrates his theory with some laboured jokes.

The best one is this example of irony:

> *The heir of a rich man complains that he cannot get the professional mourners at his relative's funeral to show suitable grief because the more he pays them to look sad, the happier they appear.*

Kant believed that convulsive laughter always required an absurd stimulus. While this is probably true, it's worth noting that it does not work in reverse. Not all absurd things are funny. M.C. Escher's etchings of impossible architecture are just as absurd as the cartoons of Rube Goldberg or Heath Robinson and yet don't provoke a laugh. The difference is that the cartoons create amusement by resolving incongruity, albeit ridiculously, while Escher's puzzles of never-ending staircases and other geometric impossibilities remain unresolved. They are like jokes with a set-up and no punchline; they go nowhere.

> *Why did the chicken cross the Möbius strip?*
> *To get to the other … no, wait …*

If you want further proof of the need for some sort of resolution for laughter to happen, consider that surrealist paintings in national art collections are not surrounded by chuckling crowds. Salvador Dalí was an admirer of the Marx Brothers and he story-boarded a script called *The Marx Brothers on Horseback Salad* that included such visual gags as an eyeball with 23 arms, 36 arms asleep on a sofa and Groucho with 6 arms answering 10 telephones. Lacking a storyline, the film would have been absurd without being funny and was never made.[23] For reason-gone-mad to be amusing it needs a narrative with a benign resolution. A heap of arms on a sofa could be either gruesome or funny, depending on how this incongruous scene is explained by the story.

Shaggy-dog stories are jokes with a long narrative set-up where the listener finally becomes the butt at the end. When I was a student in the mid-1970s, I heard the English folk singer A.L. Lloyd perform a memorable example of this kind of joke in a crowded room above a pub in Brighton:

> *In 1936 or 7, it was, there was no work to be had in London and so I signed on as a deckhand on a whaling ship called the Southern Empress. The voyage down to South Georgia in the South Atlantic took the best part of a month and at night we'd sit around and drink, sing songs, and tell stories. There was one old fella who'd been on the old sailing ships and he told us about one of his first voyages when he'd met a cushmaker. He'd never heard of a cushmaker*

before and wondered what the job might be. He watched the man day by day as they sailed towards the whaling grounds and saw that he was building some kind of structure out of wood. It took a week for him to build a box-like frame of long spars and then he began to construct a pyramid-shaped frame inside it. When that was finished, the cushmaker got a huge length of jute rope with which he wove a web suspended inside the pyramid. That took another week, and when it was done, he hauled up from the hold a large hessian sack filled with wax. It needed two men to hoist this into the structure where the cushmaker spliced it into the middle of the web. Every day the captain would ask the cushmaker, 'Are we ready yet?' and the cushmaker would reply, 'Aye, sir, we're nearly there.' After many weeks, the sense of anticipation had built to great heights and everyone on board was looking forward to the day when the cushmaker's work would be ready for use. Then, finally one day the cushmaker gave the sign and the captain ordered everyone on deck for the big event. A block and tackle was used to hoist the heavy structure up into the air and then the jib of the crane was swung out over the side of the ship. A great hush spread over the assembled men. The captain gave a nod and the cushmaker pulled on a rope that released the structure which fell with a crump into the sea.

Then, as it slowly sank beneath the waves, everyone could distinctly hear a long cussssshhhh.

When A.L. Lloyd told this story in that pub in Brighton, every step was illustrated with hand actions. He kept a roomful of people rapt right up until the final, sibilant punchline in which he spread his hands low and wide across the imaginary sea. The reaction from the room was an uproar of laughter and disbelief. It was a good story, but of course we'd all been had. It helped in the telling that A.L. Lloyd was as weather-beaten as a sailor, that he had sung us sea shanties and that he really had worked on a whaling ship.

As a scalpel applied to the body of humour, the cushmaker joke demonstrates, perhaps more clearly than most, the importance of cognition in humour. You need to pay attention to the story and allow it to build a picture in your mind, while all the time you are trying to make sense of this developing image. In the end, the elements are the same as in any joke: incongruity, absurdity, a set-up and a punchline, but the long narrative clearly requires the listener to do a lot of the work. That's why you feel duped at the end. All jokes require work on the part of the listener, but not as much as this one for so little reward. A cushmaker makes a cush. What did you expect, sucker!

In the years since 1790, philosophers and

psychologists have batted Kant's theory of humour to-and-fro, adding a qualification here, a refinement there, until producing something like a modern consensus centred around the importance of incongruity and its resolution. We can now see how this plays out in the brain. The three central processes involved are thinking (understanding), emotion (amusement), and motor control (the physical act of laughing).

Cognition starts with a set of expectations that we take from prior experience and then apply to the information provided by a set-up like '*One morning I shot an elephant in my pyjamas.*' Our initial hypothesis is that it's Groucho wearing his pyjamas. The mind is an hypothesis-generating machine and is doing this work continuously on all its sensory inputs. This process of hypothesis-generation is not confined to humour; it is as fundamental to cognition as a beating heart is to blood circulation. When you first read a sentence with a simple construction like 'Paris in the the spring', it's normal to fail to notice that it contains an error. Your brain perceives what it expects to read, not what is actually on the page. In 2018 the airline Cathay Pacific mis-spelled its own name in huge letters on the side of one of its planes.[24] No one at the company seems to have noticed, until passengers asked on social media whether it meant that Cathay Paciic no longer gave-a-F?

An hypothesis is, by definition, a tentative idea that needs to be tested against more information. Now comes the punchline with the information: '*How he got in my pyjamas, I don't know.*' This requires us to revise our initial hypothesis: So, it was the elephant wearing the pyjamas, not Groucho. The two hypotheses are incongruous and can only be resolved in an absurd world where elephants wear pyjamas. For some reason we find that funny. We'll explore why we laugh at incongruity later. In the meantime, you might be wondering whether there is any scientific evidence that it is incongruity and not say, elephants or pyjamas or Groucho's famously eccentric appearance, that makes us laugh. It's probably true that these things contribute to our sense of amusement too, but there is neurological evidence that specific areas of the brain cortex are responsible for detecting incongruity. Once detected, a different area processes the resolution (the punchline), another brain region generates the emotion of amusement and a fourth controls the muscular motions that produce laughter.

A non-invasive technique called functional Magnetic Resonance Imaging (fMRI) makes it possible to see inside someone's head to detect which particular parts of their brain are functioning while they are listening to a joke or performing a mental task. Psychologist Dr Yu-Chen Chan and her team in Taiwan used fMRI to pinpoint the different regions of the cortex involved in the cognitive processing of humour. Volunteers were presented with jokes in three different versions.[25] All versions had the same set-up, such as in this example:

Peter bought some farmland and started ploughing it with a tractor. Not long afterwards, he found a front tooth that the tractor had dug up. He felt a bit strange but kept on ploughing. About a hundred meters later, he found another tooth. 'Something is definitely wrong,' he thought to himself. After just 30 more steps, he found several more teeth. Now he was really frightened. That night he wrote to the previous owner of the land and asked, 'Was this piece of land ever used as a graveyard?'

In the first version of this joke the punchline was:

Two days later, the old owner replied, 'No. Don't worry. It used to be a football field.'

This version of the joke contains both incongruity and resolution. To tease apart the brain responses to incongruity and resolution, two other versions were also presented to the volunteers. In one of these, the punchline contained resolution but no incongruity:

Two days later, the former owner replied, 'Yes, actually it was a graveyard.'

And in the third version the punch line was nonsensical, so that although it was incongruous, there was no resolution:

Two days later, the old owner replied, 'Yes, the cliff is now behind you!'

Comparing brain scans for people receiving the different combinations of incongruity and resolution across a total of 64 different jokes revealed that incongruity is detected in two parts of the brain cortex (the right middle temporal gyrus and the right medial frontal gyrus). When the incongruity is resolved, this occurs in two other parts of the cortex (the left superior frontal gyrus and the left inferior parietal lobule). Similar experiments have identified four other brain areas — mainly in the amygdala and subcortex — that process the sense of amusement triggered by the humorous resolution of incongruity.[26]

These different parts of the brain connect in a neural circuit that runs: incongruity detection → incongruity resolution → feeling of amusement.

Neural activity then spreads to the hypothalamus and the brainstem, which operates the muscles that turn amusement into actual physical laughter. These experiments and others like them confirm that the human brain processes humour through steps that correspond to those proposed by the incongruity-resolution hypothesis. For economy I'm simply going to call this the 'incongruity hypothesis' from now on because we are going to make a bit of a song and dance about it.

Chapter Three

Song and Dance

The Philharmonic Orchestra played Beethoven last night. Beethoven lost.

This is of course a verbal joke, but you only need to hear the Portsmouth Sinfonia, possibly the world's worst orchestra ever to cut a disc, ponderously sawing their way through the bars of the *William Tell Overture* to appreciate what musical incongruity sounds like and how hilarious it can be.[1] But, could it be superiority as well as incongruity making us laugh? Maybe we laugh at bad musicians as well as bad music? Possibly so in the case of the Portsmouth Sinfonia, but the superiority hypothesis cannot explain why the funniest musical jokes are made by the best musicians, not the worst ones. Only the incongruity hypothesis can explain that. In the style of musical composition called a *quodlibet*, the composer juxtaposes melodies from different sources, surprising the listener to comic effect. Spontaneous quodlibets were performed at reunions of the Bach family when, according to J.S. Bach's biographer:

> As soon as they were assembled, a chorale was first struck up. From this devout beginning they proceeded to jokes which were frequently in strong contrast. That is, they then sang popular songs partly of comic and also partly of indecent content, all mixed together on the spur of the moment … This kind of improvised harmonizing they called a Quodlibet, and not only could [they] laugh over it quite whole-heartedly themselves, but [it] also aroused just as hearty and irresistible laughter in all who heard them.[2]

The family tradition even crept into J.S. Bach's day job. The last (30th) variation of the *Goldberg Variations* for piano is a quodlibet that quotes various German folk songs, including one called '*Cabbage and turnips have driven me away, had my mother cooked meat, I'd have opted to stay.*' Many other classical composers made musical jokes, including C.P.E. Bach, Debussy, Haydn, Mozart and Tchaikovsky.[3]

In modern times, the great exponent of the quodlibet is the musical humourist Peter Schickele (*b*.1935) who would perform the newly discovered works of P.D.Q. Bach (1807–1742), the 21st of Johann Sebastian and Anna Magdalena Bach's 20 children. P.D.Q. Bach's works were all recorded in front of a live audience and the recordings have been analysed to diagnose the musical causes of the hilarity at Schickele's shows. Schickele's (aka P.D.Q. Bach's) musical jokes include abruptly switching musical genres, such as in his 'Unbegun' Symphony where he suddenly inserts a trumpet playing 'Ta-ra-ra-boom-tee-eh' and 'De Campdown Races' into a lyrical andante. In total there are nine different kinds of musical gag in the oeuvre, and they all work by violating the expectations of the audience with an incongruity of some kind. In the recordings, the switching-genres gags got the biggest laughs, as was the case back in the days of Johann Sebastian.[4]

Scans of the auditory cortex of the brain show that people listening to a familiar melody are very quick to detect a note in the wrong key, responding to this incongruity in just 1/10th of a second.[5] In fact, people familiar with western music, but who have no formal musical training, can detect incongruities even in melodies that they have never heard before.[6] Though this brain response has so far only been studied in people exposed to the western musical tradition, there is every reason to suppose that it operates in other cultures as well. For example, musical jokes are common in Javanese gamelan music, where musicians introduce incongruities into otherwise highly structured compositions for their comic and dramatic effect.[7]

There is evidence that even outside the arena of language and music, incongruity can be amusing. In an ingenious experiment performed 50 years ago, Swedish psychologist Göran Nerhardt presented student subjects with a simple task designed to test their reaction to incongruity in a joke-free situation.[8] Each subject was presented with a series of weights and as they were handed each one, they were asked to judge how heavy or light it was on a 6-point scale from very light (actual weight 740g) to very heavy (2.7 kg). After many trials sampling the full range of weights, the subject was finally presented with a weight much lighter than any s/he had lifted before. Being unexpectedly presented with an incongruously light weight stimulated laughter.

This experiment, and subsequent ones like it, support the idea that incongruity in non-threatening situations is itself funny, even if removed from the context of jokes and overt humour. However, there's a feature of the Swedish study that suggests we need to qualify how it is interpreted. Nerhardt first tried out his experiment with passengers waiting at a railway station, but it failed. It seems that in this situation, people just weren't in the mood to laugh. In complete contrast, the students in the subsequent successful study were a hilarious lot and laughed quite a bit during the experiment, even when not handling incongruous weights. Incongruity increased their laughter, but the result was not all-or-nothing. So, exactly as Charles Darwin observed a century earlier, while we can conclude from Nerhardt's study that incongruity *per se* can trigger laughter, you have to be in the right mood and not anxious about missing a train.

The discovery that all kinds of incongruity can tickle the funny bone has deep implications for its evolutionary origin. Though jokes and conversational banter are the dominant modes of humour in our lives now, the language skills needed for spoken humour is comparatively recently evolved, possibly in the last half a million years.[9] Verbal humour must therefore surely have emerged long after the more general humorous response to incongruity.

The first evolutionary step towards humour must have begun with a general mental ability to compare expectations with various sensory inputs, including those from vision, hearing and touch. This ability would have been vital to survival. From there, incongruities that could be benignly resolved became a trigger for laughter. Much, much later, when language evolved, the incongruities contained in speech became attached to the existing humour mechanism and *voila!* The joke was born.

The incongruity hypothesis explains many of the properties of humour. First, it explains why humour can be so subjective. What might seem funny to one person may leave another struggling to see the joke because they do not have the same expectations and experience. In his experiments with weights, Nerhardt had to establish some expectations in his subjects about the range of weights he was using before presenting them with an incongruous one that would make them laugh. It is said that Sir Isaac Newton, who was better known for gravity than levity, only laughed once in his entire life, when someone asked him what use he saw in Euclid's *Elements*. In his mind that was an absurd question because he had been studying Euclid's geometry since childhood. To most of us ordinary mortals, the question might make us think, or leave us puzzled, but not make us laugh.

The subjective nature of incongruity in humour can also explain why superiority features in so many jokes. In jokes made at someone else's expense we are laughing at the difference (incongruity) between them and us. How can David Levi, or whoever, be so stupid? These jokes work if you know who David Levi is or if you share the prejudice that the Irish/Poles/Sikhs are stupid. If you don't, you may understand the joke on an intellectual level, but not find it funny. The reason superiority occurs so frequently in jokes is just because differences between 'us' and 'them' are such a rich and shareable source of incongruity. So, from a theoretical point of view, anything that the superiority hypothesis can explain can also be explained by the incongruity hypothesis. This makes the superiority hypothesis redundant. The incongruity hypothesis grasps and explains more of the elephant that is humour.

The incongruity hypothesis can also account for why explanations ruin jokes. An explanation pricks the balloon by turning mad reason into just plain reason. To be funny, incongruity needs to be resolved by colliding two incompatible interpretations, but by explaining a joke you are making the different interpretations compatible and so the humour evaporates. Repeating a joke can have the same effect because once we have heard it, the incongruity becomes part of what we expect from the world. Hence Hobbes' idea that jokes must contain novelty and surprise.

A man goes to prison and the first night he's trying to get to sleep when he hears a prisoner yell out, '41!' followed by a chuckle from his cellmate. He thought nothing of it, but then there was a cry of '33!' and another chuckle.

'What's going on?' he asks his cellmate.

'Well, we've heard every joke in here so often, we've numbered them to save time.'

'Oh,' he says, 'can I give it a try?'

'Sure, go right ahead.'

So, he yells out '102!' and there is uproar. Hysterical laughter sweeps from cell to cell and landing to landing. Eventually the laughter subsides, and the newbie turns to his cellmate who is wiping tears of mirth from his eyes. 'That was a good one, eh?'

'Yeah! We ain't never heard that one before!'

Novelty is good, but children love to hear and re-tell familiar jokes and there are some jokes that adults want to hear again and again. This is difficult to explain with the incongruity hypothesis alone: something more is needed. I suspect that in these cases, the joke is not just a joke but a means of bonding and re-visiting a moment of remembered enjoyment. Don't forget that laughter occurs when we are happy, not just when we hear something funny. Ken Dodd was constantly asked by audiences to repeat a joke about a three-legged chicken because of the engaging way in which he performed it. I won't spoil it for you.

Look up the video online.[10] The first time you see this joke performed the incongruity will make you laugh. Thereafter, repetition will replay the laughter from happiness.

There are other challenges to the incongruity hypothesis. For example, philosopher Steven Gimbel has argued that songs like Tom Lehrer's 'When You are Old and Gray' make us laugh because of the very congruence of the rhyme, which ingeniously lists all the reasons Tom's girlfriend should not delay making love with him, because when they are old:

An awful debility,
A lessened utility,
A loss of mobility
Is a strong possibility.
In all probability
I'll lose my virility …

… and so on for another ten lines of rhyming ingenuity.

I'd say that sustaining a rhyme like this is pretty incongruous, unexpected and unusual, but this is a subjective judgement on my part. And there lies the problem: the subjectivity of deciding what is incongruous and what is not. How can the incongruity hypothesis be called scientific if it relies on a concept that is too vague for it to be tested? Maybe incongruity only appears to be necessary to humour because it is confounded with a multitude of other things. Steven Gimbel thinks we laugh at 'When You are Old and Gray' because it is so clever.[11] Jokes often contain superiority, absurdity and references to bodily functions that make us laugh. When I was a student, some friends and I produced a magazine and on the back cover we printed illustrated instructions on how to make origami contraceptives. We felt ourselves to be very clever and the joke had our sense of superiority, plus absurdity and sex all rolled into one. Can incongruity be teased out and isolated from such a confection?

A trio of psychologists took up this challenge by setting 300 student volunteers the task of generating humour with pairs of words that varied in incongruity and reference to sex.[12] For example, students were given the unrelated and therefore incongruous word pair *money* and *chocolate* or the pair *money* and *sex* and asked to list five ways in which they were similar or different. The related (congruent) word pairs for comparison were *love* and *friendship* and *love* and *sex*. Responses included such gems as:

> *The difference between money and chocolate is that one swells the wallet and the other swells the hips.*

The similarity between money and chocolate is that neither lasts very long.

The similarity between love and friendship is that both have the letter 'e'.

The funniness of all the responses was evaluated by independent judges who had not participated in the task. The study found that incongruous word pairs produced funnier responses than congruent pairs and that responses highlighting differences were funnier than those highlighting similarities. The presence of sex or other emotionally loaded words had no significant effect on how funny the responses were. In this study it was incongruity and incongruity alone that made the responses funny.

Incongruity is not only a necessary element of humour, but it can be used to recognise a joke.

A rabbi, a priest, and a minister walk into a bar. The bartender says, 'What is this — some kind of joke?'

The bartender recognises a classic joke set-up — and that's the joke. Even gorillas have heard these jokes:

A rabbi, a priest, and a gorilla walk into a bar. The gorilla looks around and says, 'I must be in the wrong joke.'

These are jokes about jokes. If you were one of those sad people who take humour too seriously, you'd label such things meta-jokes and keep them secretly hidden away under your bed in a velvet-lined box where nobody could laugh at them. Occasionally, perhaps on birthdays, you'd take them out and gloat.

But the bartender's question is also a research problem in computational linguine. I mean computational linguistics. This is, confusingly, not a field of linguistics but a branch of computer science that aims to write software that can process language the way humans do.[13] Surely, it should be possible to get a robot bartender running such software to recognise that if three clergymen walk in together, this scenario presages a particular kind of joke? I'm afraid not.

A piece of string walks into a bar and asks for a martini. 'I'm sorry, but we don't serve pieces of string,' says the bartender.

'But, you can serve me,' replies the string.

'Why, aren't you a piece of string?' asks the barman.

'No. I'm a frayed knot.'

The problem for a robot is that sometimes string is just string, and sometimes it's knot. Three computational linguists called Strapparava, Stock, and Mihalcea walked into a bar and dreamt up a way to train a computer how to recognise a joke.[14] They fed a computer running machine-learning software thousands of one-liners like these:

Take my advice; I don't use it anyway.

I get enough exercise just pushing my luck.

Beauty is in the eye of the beer holder.

The software was also fed with statements of similar word length that had no humorous content. Having been instructed which statements were funny and which were not, the software was tested to see whether it could distinguish new one-liners from non-humorous text it had also not seen before. The computer was good at distinguishing drollery from dross, but not always. When fed a mixture of one-liners and snippets of news reports, the software spotted the joke 76 per cent of the time. However, it only got the joke 53 per cent of the time when fed a mixture of one-liners and proverbs. This is not very impressive, since by chance alone even a dumb slug would pick the right answer out of a choice of two 50 per cent of the time.

It turned out that the machine-learning software

had not learned how to spot jokes from their incongruity at all, but had instead homed in on the fact that a lot of one-liners contain specific kinds of linguistic structure.

Such as alliteration:

Infants don't enjoy infancy like adults do adultery.

Antonyms:

Always try to be modest and be proud of it!

Or key words such as 'sex' or 'behind'. The software got really excited when all three types of clue occurred in a single piece of text, such as:

Behind every great man is a great woman, and behind every great woman is some guy staring at her behind!

Not pretty. Proverbs also often contain alliteration, antonyms and slang, so of course the poor software had trouble telling a proverb from a one-liner. Proverbs can readily become jokes:

Familiarity breeds contempt — and children. (Mark Twain)

Red sky at night: shepherd's delight. Blue sky at night: day. (Tom Parry)[15]

An apple a day keeps the doctor away. An onion a day should take care of everyone else.

On the other hand, news reports, at least those from Reuters, where the researchers sourced their material, are not distinguished by the use of alliteration, antonyms and slang, so the software was much better at spotting a joke in amongst the morning news. And this seems to be where the three researchers, like so many others before them, decided to leave the humour field for ever.

As you might imagine, if it has so far proved impossible to get a computer to recognise incongruity in humour, we should not expect them to be great jokesmiths either. Could this be a robot joke?

A robot walks into a bar. 'What can I get you?' the bartender asks.

'I need something to loosen up,' the robot replies. So, the bartender serves him a screwdriver.

Nope. Robot-generated jokes tend to be even more feeble than this one. Here is the kind of thing that Apple's OS9 speech system could knock out if you asked it to tell you a joke in 1999[16]:

You: Computer, tell me a joke.
Computer: Knock, knock.
You: Who's there?
Computer: Thistle.
You: Thistle who?
Computer: Thistle be my last knock knock joke.

Promises, promises. Twenty years later, computer-generated jokes are still puns manufactured from lists of homophones like:

What kind of temperature is a son? A boy-ling point.

What kind of tree is nauseated? A syc-amore.[17]

To see how far short these jokes fall from the best that humans can do, compare them with a random selection of punning definitions from the spoof *Complete Uxbridge English Dictionary:*

Agog: A half-built Jewish place of worship.

Crucifix: Religious adhesive.

Elfish: Spanish seafood.

Gastronomy: The study of Michelin stars.

Zebra: The largest size of support garment.[18]

Linguistics is the science that seeks the deep, underlying structure of language. It's a hard row to hoe, full of contrary data. In a lecture at Columbia in New York, an eminent Oxford professor of linguistics once explained to his audience that, while in many languages a double negative was used to convey a positive (e.g. 'She is not unlike her brother' meaning 'she is like her brother'), there was no language in which a double positive implied a negative. From the back of the hall came the sarcastic reply 'Yeah, yeah.'

This killer riposte came from Sidney Morgenbesser. His quick wits were schooled on the streets of New York's Lower East Side and he trained to be a rabbi before switching to more secular arguing.

Stories about Morgenbesser's wit are many.[19] Enduring a long illness near the end of his life, he enquired of a fellow philosopher at Columbia:

> *'Why is God making me suffer so much? Just because I don't believe in him?'*[20]

Rationalists are generally happy to compromise if there's a joke to be made. After all, they already know they won't be going to heaven:

Thank God I'm an atheist. (Luis Buñuel)

Before the Second World War, Nobel Prize-winning physicist Niels Bohr was entertaining a visiting scientist at his lab in Denmark. The visitor expressed surprise at seeing a lucky horseshoe nailed over the door. 'You aren't superstitious, are you?'

'Oh, no,' said Bohr, 'but they tell me it works, even if you don't believe in it.'

Arthur C. Clarke's version of the same joke was:

'I don't believe in astrology. I'm a Sagittarian and we're sceptical.'

These jokes are mad reason perfected. They illustrate why making good jokes is a much more difficult problem for artificial intelligence (AI) than relatively simple things such as beating all-comers at chess and Go — two milestones that computers have already accomplished. Humour is the kind of problem that cognitive scientists working in AI call AI-complete, meaning that you need to get a machine to think like a human before you can expect it to be funny.[21] To solve an AI-complete problem, a computer must pass the so-called Turing Test, which means it must be indistinguishable from a human in a disembodied conversation such as an exchange of text messages. However, not all cognitive scientists agree about this test. Marvin Minsky, one of the founding fathers of AI, described the Turing Test itself as a joke. Maybe a better test of AI-completeness would be the one that comedian Ken Dodd said proved Freud wrong:

> *'Freud's theory was that when a joke opens a window and all those bats and bogeymen fly out, you get a marvellous feeling of relief and elation. The trouble with Freud is that he never had to play the old Glasgow Empire on a Saturday night after Rangers and Celtic [football teams] had both lost.'* [22]

Freud's theories do seem to be more the product of bourgeois Vienna than working-class Glasgow. Likewise, computer humour is not yet ready for a live Glasgow audience.

So, what does the theory of humour amount to? Finding something funny is a cognitive response to the benign resolution of incongruity. It's almost exclusively found in communication between people (including through books), often but not always exploiting a sense of superiority over others. You need a brain to tell a joke or laugh at one — a machine simply won't do.

Not all brains are alike, of course, so how does the

variety affect what we think is funny? Psychologists, having tidy minds, like to separate their subjects into personality types and since the 1990s there has been a consensus that the essential personality differences between people may be defined by five big factors. The Big Five are: Extraversion, Emotional Stability, Agreeableness, Conscientiousness, and Openness to Experience.[23] The details of these five aspects of personality and how they were distilled from a library of studies in a babel of tongues is perhaps best left, like knowledge of what goes into sausages, to those with strong stomachs. But humour is one of the diagnostic characteristics of extraversion, so one might expect some association between personality and what someone finds funny. Such a connection is supported by an fMRI study, albeit one containing only a small number of subjects.[24] As the German poet Goethe said, in an age when women were ignored in intellectual discourse: 'Men show their character in nothing more clearly than by what they think laughable.'

Watching psychologists trying to prove this proposition with various psychometric tests is like watching small boys trying to catch the wind in a pond net. The *Mirth Response Test*, the *IPAT Humour Test of Personality*, the *3 WD humor test*, the *Escala de Apreciación del Humor*, and a host of others swipe frantically at jokes and personality, but never quite seem to capture them in close humorous embrace.[25] The evidence that the kind of humour people enjoy is related to their personality type is weak. People have different personalities, and jokes vary from the tame to the lewd and the benign to the vicious, but there seems to be very little accounting for who finds what funny. Even Christians and atheists laugh at the same things most of the time, though fundamentalists do tend to be po-faced at jokes.[26] What does shine through is the deep and universal importance of incongruity resolution.

Despite the wealth of evidence that now supports the incongruity hypothesis, there are still a few people who think they can do better. In 2017, two Australian scientists sought to use the mathematics of quantum theory to derive a new theory of humour.[27] If quantum theory can resolve the wave/particle duality of light, why not the sense/nonsense duality of humour? But even if there is an equation that can define the essence of all humour, it won't tell us why we react with laughter or why the emotion evoked by humour is pleasure rather than pain. Defining the stimulus is one thing, explaining why it provokes the response that it does is something else entirely and to do that we need to look into evolution. In the beginning was the tickle.

Chapter Four

Tickle and Play

Charles Darwin was a compassionate family man who experimented on his infant children. Shocking? Not really, as all he did was tickle them and observe that the capacity for laughter appears at a very young age. He included his observations in a substantial book called *The Expression of the Emotions in Man and Animals*, in which he demonstrated that the way we express our feelings, including laughter, is often animal-like.[1] Do animals — in particular chimps and gorillas, our nearest living relatives — laugh? Mark Twain expressed an opinion that was common in the 19th century with the one-liner:

> Man is the only animal that laughs, or needs to.

Darwin believed otherwise, based mainly on observations made in zoos and of domesticated animals. These days, we have recordings of the great apes made in the wild. These show that when tickled, orangs, chimps, bonobos and gorillas all laugh, or emit a 'play vocalisation', as psychologists call it.[2] Infant chimps engage in tickle play from just 4½ weeks old and develop characteristic gestures that invite tickling from their playmates.[3] Chimps have a play face that they use in rough-and-tumble play, though their laugh sounds very different to ours. A chimp laugh is made on the in-breath *ah-ah-ah*, in contrast to our own which is made on the out-breath *ha-ha-ha.*

Despite the unique means by which human laughter is produced, a fascinating study has shown how similar spontaneous human laughter is to animal vocalisation. Slowed-down recordings of spontaneous human laughter were played to listeners who were asked to say what they thought the sound was. Most participants identified the sounds they heard as laughter-like, but they thought they were listening to animal calls, not to humans.[4] When people were played slowed-down recordings of intentionally produced (volitional) laughter, they correctly identified the source as human. Spontaneous laughter is a call from our deep animal nature. Volitional laughter is more like speech and therefore more recognisably human.

This distinction applies to our ability to identify individuals by their laughter as well. Experiments have found that people have no difficulty recognising who is laughing from their volitional, speech-like laughter, but cannot do so as easily from spontaneous laughter.[5]

Why do animals vocalise when playing? The likely answer is that, since play is a social interaction, the vocalisation is a signal to playmates that your intentions in tickling or chasing are not threatening. This function of laughter in play may be why laughter is contagious, because the recipient of the signal must agree and reply, 'I'm playing too.' Repeating the laugh you have just heard enables you to join in the fun. The evolutionary origin of animal signals can frequently be found in earlier behaviours that have a different, but related function.[6] In the case of play vocalisation, the earlier function may have been an all-clear signal given to family members when a threat, a predator for example, has gone away.[7]

Chimps and humans share a common ancestor who lived about 6.5 million years ago. That ancestor likely had a play vocalisation too, but it was probably much more like a chimp's than our own, since our laugh is shaped by the much more recent evolution of speech. When speech evolved in humans, evolution attached a new trigger to the play vocalisation of the speaking primate: verbal humour. In Charles Darwin's words, 'The imagination is sometimes said to be tickled by a ludicrous idea; and this so-called tickling of the mind is curiously analogous with that of the body.'[8] How did evolution make us react to the resolution of incongruity as though we are being tickled?

Evolution travels in small steps, always starting from what already exists.

A foreign tourist in Galway asks one of the locals for directions to Dublin. The Irishman replies: 'Well, if I were you, I wouldn't start from here.'

Evolution always starts from wherever 'here' is and what is more, it never asks for directions. Consequently, many of the features of humorous laughter, such as the pleasure it evokes, its social nature and contagiousness, are there in its ticklish and playful origins.

Aristotle noticed that one cannot tickle oneself and he speculated that to respond to tickling you perhaps need to be taken by surprise.[9] The ineffectiveness of self-tickling is indeed strange since it is possible to induce other pleasurable sensations with your own hand, or at least so I am told. Playing with others can involve tickling, playing with oneself cannot. Is evolution trying to tell us something?

Two thousand years later, a trio of psychologists and a robot got around to testing Aristotle's hypothesis with an experiment in which experimental subjects tickled themselves with a robot arm.[10] Subjects controlled the tickling robot with their left hand and received a tactile stimulus from the robot to their right hand. By introducing a variable time delay between the trigger from the left hand and the stimulus to the right, the experimenters were able to test whether being able to tickle yourself is just a matter of time, rather than surprise. It turned out that delays of less than 1/5th of a second were sufficient to make the tactile stimulus from the self-controlled robot feel ticklish, even though most subjects were unaware that there was actually any delay at all. Finally, a robot that can make people laugh!

Comedians know very well that performance is all a matter of timing, but something else entirely was going on in this robot experiment. In order to control the actions of our muscles, the brain not only initiates movement, but also keeps track of its commands in a kind of virtual model of the body. This is what stops you being startled by your own actions as well as tickling yourself. The model is continuously updated to keep track of movements, so even a very short delay is sufficient for the inhibition caused by the movement of the left hand to die away, allowing the right hand to feel tickled.

The virtual body model that stops you tickling yourself provides a means of distinguishing self from other, and thus is a part of the mental construction of the self. Psychiatric patients who hear voices, or who have the sensation that their actions are not under their own control, feel just as tickled when they apply a stimulus with their own hand as when they are tickled by another person.[11] This is the benign consequence of a more general psychological impairment to the ability to fully distinguish self and non-self.

When I'm not in a relationship, I shave one leg so it feels like I'm sleeping with a woman.[12]

One way or another, tickling and the laughter it generates are playful. While most of us cannot tickle ourselves, we are very particular about who we allow to tickle us. Non-consensual tickling feels like an assault. Robert Provine found in a survey of more than 400 people of all ages that tickling is practically confined to interactions between friends, family and lovers and that it is usually reciprocal.[13] From adolescence onwards, tickling predominantly occurs between the sexes and in adults it is a part of foreplay. Here too, we see a characteristic of tickling that is a forerunner of the role that humour plays in courtship.

In early life, play is necessary for youngsters to learn how to interact safely within the group. From an evolutionary perspective, it makes sense that tickling, which is something our neural wiring prevents us doing for ourselves, became the evolutionary basis for socialisation through play. Laughter is the universally understood response that signals 'I like it!'

Play is instinctual in young mammals, not just in primates, and it is anciently evolved. Tickle a rat and it will emit an ultrasonic laugh at a pitch of 50 kHz, way above what we can hear.[14] Rat laughter, like the human kind, is infectious and young rats will choose to spend time with adults that laugh more in preference to those that laugh less.[15]

At what frequency does laughter become painful? 1 Gigglehurtz.

In a recent study that must be a shoo-in for a future Ig Nobel Prize, a team of psychologists in Germany played hide-and-seek with rats in a large room equipped with hiding places for humans and rodents.[16] The rats took about two weeks to learn the rules of the game, being rewarded with a tickle when they hid and were found, or when they searched and found the experimenter. The most extraordinary finding was that rats playing the game vocalised ultrasonically throughout, except when they were hiding. Neural recordings made in the rat brain showed that they were using different areas of the prefrontal cortex depending on whether they were hiding or seeking. This suggested that the rats mentally distinguished between different roles when playing.

The next time you frolic with your dog, think of the common mammalian ancestor of dogs, humans, and rats that lived 96 million years ago and from which we all inherited the capacity to play. Of canine laughter, Max Eastman (1883–1969)[17] observed:

> Dogs laugh, but they laugh with their tails. What puts man in a higher state of evolution is that he has got his laugh on the right end.

A more scientifically minded study of dogs found that they do display characteristic facial expressions when playing, but that the muscles involved in a dog's happy-face are different ones to those used in a human face displaying the same emotion.[18] This is not a trivial conclusion, since it means that play and signalling emotional state to playmates has evolved in a wide range of animals, but that this is achieved by different means in different species. In other words, what is universally important is signalling non-hostile intent, rather than the precise means of doing this. We see the same principle at work in the cultural evolution of human language. There is a diversity of words for saying the same things that people want to talk about.

An Irish professor of languages is chatting to a Spanish colleague who asks him whether there is an Irish equivalent of 'manaña'. The Irishman thinks for a moment and replies, 'Yes, but it doesn't convey the same sense of urgency.'

Not to be outdone by their zoological brethren, botanists have tickled plants to see what happens. They found that gently stroking the leaves of young thale cress produced chemical changes in the plant that delayed sexual maturity and flowering.[19] Whether repeated tickling could keep plants in a state of perpetual, playful youth we do not know. I could write a whole book about what we do not know about plants. Actually, I know so little that it would be a very long book indeed. But I digress.

We need to talk about pleasure. Play is a pleasure and if Darwin was right that humour is a tickling of the mind, then the pleasure of humour came as a part of the package when humour and laughter were hooked together by evolution. Could the association of pleasure with ticklish laughter be not just the source of humour's pleasure, but the very advantage that brought them together and made humour funny?

What good is pleasure to evolution, anyway? Moralists would have us believe that pleasure is what leads us astray. Lust, gluttony and sloth are the vices that result from sexual pleasure, appetite and reserving one's energy for sex and eating.

> *Moses comes down from Mount Sinai carrying the two tablets of stone that God has given him. 'My people,' he says, 'I have good news and I have bad news. The good news is that I have got it down to just ten commandments. The bad news is that adultery is still in there.'*

Biologists have a quite different perspective: pleasure is fundamentally good for you. In fact, the simplest idea about the evolution of humour that anyone has come up with is that it evolved because it allows us to enjoy ourselves more.[20] But there is a problem with this hypothesis because pleasure in itself provides no evolutionary advantage. Possessing a hypothetical gene for enjoyment will not increase the number of offspring you have, unlike, for example, a liking for sex. It's obvious how natural selection could favour the spread of a gene that makes sex fun. But why would natural selection favour a general capacity for enjoyment? So, plausible though the enjoyment hypothesis may at first sound, it does not really explain anything.

A better evolutionary hypothesis is that pleasure exists because it directs our behaviour towards actions that enable us to survive and to reproduce.

The pleasure of sex is the most obvious example, but our choice of nutritious food is equally influenced by evolutionary imperatives.

Now that food has replaced sex in my life, I can't even get into my own pants!

There are specific sensors on the tongue that detect the presence of sugar, a breakdown product of starch, sending the brain a signal that we perceive as pleasurably sweet — and carbohydrates like starch supply the energy we need. Another kind of sensor detects glutamate, a component of proteins, and this we experience as the pleasurably savoury taste called umami — proteins are essential to health.

There are other taste receptors that detect dietary errors, such as bitter foods that may be poisonous. There are two kinds of response to salt, which is tasty at low concentration but distasteful when dangerously high. These links between food stimulus and taste response are wired-in. This has been demonstrated by using genetic techniques experimentally to rewire a rat's taste buds, causing it to eat salt as though it were sugar.[21] A joke in poor taste.

Recognising incongruity is a much more complicated trick for the brain to pull off than sensing the presence of sugar on the tongue, so the wiring is a good deal more complicated, but ultimately the wires end up in the same place — stimulating pleasure in the amygdala. Dr Samuel Johnson once said, 'Pleasure is very seldom found where it is sought,' but then he was a doctor of letters and was probably looking in the wrong drawer. Nonetheless, pleasure is a complicated emotion that is influenced by many pre-disposing factors, both inside and outside the brain. If you are not in the mood for humour, nothing will make you laugh. On the other hand, if you are in a laughing crowd, you could find yourself laughing without even having heard the joke.

Apes play and enjoy a tickle, but can they recognise incongruity and take a joke? Abe Goldberg did some fieldwork at the Bronx zoo to find out:

Abe went to the zoo one day. While he was standing in front of the gorilla's enclosure, he noticed the gorilla watching him intently. The man waved at the gorilla, the gorilla waved back. He patted his stomach and the gorilla copied him. He jumped up and down, the gorilla started jumping. He made faces, pulled his hair, hopped on one foot, spun in a circle, and beat on his chest. His antics were copied exactly by the gorilla in the cage.

All of a sudden, the wind gusted and he got some grit in his eye. Abe rubbed his eye, trying to make it better. While doing so, he stepped closer to the cage. As he pulled his eyelid down to dislodge the particle, the gorilla went crazy, banged against the bars, reached out, grabbed the nearly blinded man, and beat him senseless. When the zookeeper heard the commotion and came over, Abe told the keeper what had happened. The zookeeper nodded and explained that in gorilla language pulling down your eyelid means 'fuck you'.

The explanation didn't make the gorilla's victim feel any better, but he accepted it. As he left, he became madder and madder. He plotted his revenge. The next day he purchased two large knives, two party hats, two party horns, and a large sausage. Putting the sausage in his pants, he hurried to the zoo and over to the gorilla's cage, into which he tossed a hat, a knife, and a party horn.

Knowing that the big ape liked to mimic people, he put on a party hat. The gorilla looked at him, and looked at the hat, and put it on. Next he picked up his horn and blew on it. The gorilla picked up his horn and did the same. He twirled in a circle blowing the horn. The gorilla did the same. Then Goldberg picked up his knife and waved it over his head. Again, the gorilla copied it. Next, he whipped the sausage out of his pants, and sliced it neatly in two. The gorilla looked at the knife in his big hairy hand, looked at his own crotch, and pulled down his eyelid.

What is surprising about this joke is that most of this gorilla's behaviour has actually been observed in captivity, though apes tend not to be allowed to beat up visitors these days. Mimicry is fundamental to play and to one of its chief evolutionary functions: learning, though there was not much that Abe could teach his gorilla.

Although gorillas and chimps are not capable of vocalising like humans, some living in captivity have been taught a modified version of American Sign Language (ASL), enabling them to converse with their carers. The most celebrated of these is a female gorilla called Koko, whose training began when she was a year old. When she was five and a half, she was given an intelligence test designed for children and achieved a score equivalent to a child of nearly five. Koko died in 2018, but during her 46 years she acquired an active vocabulary of a thousand signs and could apparently understand more than two thousand words of spoken English.[22]

Koko had a sense of humour remarkably like that of a child of five. She loved to be tickled and on YouTube you can see her playing with the late comedian Robin Williams who is cracking up with laughter himself.[23] Koko would play games of pretend, for example taking a rubber tube and signing that she was an elephant and wanted to drink her favourite fruit juice through her elephant nose. She'd also tease people by offering them inedible objects to see if they would eat them. Like a toddler, Koko would deliberately call things by the wrong names, such as in this dialogue where her carer, Barbara (B), showed her a picture of a bird in a magazine:

K: THAT ME.
B: Is that really you?
K: KOKO GOOD BIRD.
B: I thought you were a gorilla.
K: KOKO BIRD.
B: You sure?
K: KOKO GOOD THAT (pointing to the bird).
B: OK, I must be a gorilla.
K: LIP BIRD YOU. ('LIP' is a sign that Koko has consistently used when referring to or naming a human female.)
B: We're both birds?
K: GOOD.
B: Can you fly?
K: GOOD. ('Good' can mean 'yes'.)
B: Show me.

> K: FAKE BIRD CLOWN. (She laughs.)
> B: You're teasing me. (Koko laughs.) What are you really?
> K: (Koko laughs again and after a minute signs:) GORILLA KOKO.[24]

Koko's laughter seems to reveal the knowing silliness, or incongruity, of her pretending to be a bird. She was once explicitly asked what she thought was funny, and she signed HAT, pointing to a rubber key that she had previously placed on her head, pretending it was a hat. Another carer who put a peanut shell on her own head and called it a silly hat got a playful response from Koko who pulled down her lower eyelids and stuck out her tongue.

Young children enjoy simple play on words and so did Koko, for example signing NEED on her knee, apparently conscious that the two different words sound alike and making a pun. She would also play pranks, once tying her carer's shoelaces together and signing CHASE when they were playing with each other. But, how much did Koko really understand? Sceptics pointed out that sometimes Koko's carers asked leading questions and that their interpretation of her responses were not objective. She was prone to sign NIPPLE at random, and would be told not to be silly and to try again.[25] Was this childlike mischievousness, or simian incomprehension?

When The Gorilla Foundation staged an online chat between Koko and the public, aided by human interpreters, this presented an open goal for satirists and *Scientific American* published 'the lost Koko transcripts':

> Q: What bothers you?
> Koko: Grad students. I am not an animal. Well, you know what I mean.
> Q: Is signing hard to learn?
> Koko: I continue to confuse 'heuristic' with 'hermeneutic'.
> Q: Can you read?
> Koko: I find Woody Allen's early writings piquant. Hemingway used little words to say big things. I've dabbled in Chomsky but find him pedantic, and I disagree with fundamental aspects of his theses. Goodall raises some interesting issues.
> Q: Where does a big gorilla like you sleep, anyway?
> Koko: Wait for it ... anywhere I want. Of course.[26]

Normally, science requires more than a single instance of a phenomenon to establish that it is real, but Koko's behaviour is so well recorded and witnessed by so many people during her lifetime that it would be difficult to deny the evidence she has provided that gorillas are capable of humour and even practical jokes. What is more, chimps show many of the same behaviours shown by Koko and seem particularly fond of jokes that involve urinating on their carers and flinging faeces at zoo audiences who usually also think the joke is hilarious. And why wouldn't they since, after all, we too are great apes.

Do apes display the same jocularity in the wild? Perhaps gorillas sit around and swap jokes and toilet humour that would make a toddler squeal with delight? Probably not, but Koko and a male gorilla who was also taught to sign were observed to converse with each other without human mediation, so it's not impossible that if gorillas are saved from the extinction that threatens them in the wild, they will have something to say to each other about it. 'Close shave', perhaps? Observation of wild chimps has recently revealed that they do have their own language of gestures, using signs that apparently mean things like *stop that, move away, give me that, follow me, let's groom*, and inevitably, *let's fuck.*[27] Sadly, though, bands of chimps that live near humans in the wild lead disrupted social lives and have an impoverished behavioural repertoire.[28] Could we be underestimating chimp capacity for humour as a result?

Whether or not apes use humour in the wild, they do play and it's still important from an evolutionary perspective to know that our fellow apes are at least capable of recognising incongruity and laughing. This is because knowing that our nearest living primate relatives have the mental capacity for humour tells us that this trait in us must be ancient. When related species share common characteristics, the simplest interpretation is that they all inherited this from a common ancestor. Humans and chimps shared a common ancestor about 6.5 million years ago and that ancestor parted evolutionary company with its common ancestor with gorillas another 2.5 million years before that.[29] If chimps and gorillas behave and think in some respect like we do, we are almost certainly looking at mental faculties in ourselves that evolved more than 9 million years ago.

Evolutionary psychologist Robin Dunbar has suggested that laughter played an important part in what happened in human evolution after we split from the common ancestor with chimps. Dunbar has proposed that the advantage that drove the evolution of laughter in our ancestors was that laughter creates a kind of glue that binds social groups together.[30] Many animals live in social groups that offer individuals protection from predators, mating opportunities and, in the case of pack-hunting animals, access to food. Primate societies are particularly tight-knit and are held together by long-lasting affiliations between individuals who consistently groom each other. According to Dunbar, grooming is too time-consuming to allow groups bigger than 50 individuals to bond, so as human societies grew beyond that size, there would have been a need for an alternative to hold them together. Laughter, which Dunbar argues is a kind of vocal grooming, was that glue and could have operated before the evolution of speech.

There is little doubt that Dunbar is right about the cohesive effect of shared laughter and we shall look at the evidence for this later. Social laughter triggers the brain to release endorphins, which are the body's own versions of the drug morphine.[31] Endorphins produce a general feeling of well-being, they increase emotional attachment and create a greater receptiveness to the next gag. Victor Borge, once dubbed the funniest man in the world by *The New York Times*, put the result succinctly when he said that 'Laughter is the shortest distance between two people.'

However, the limitation of Dunbar's hypothesis is that for laughter to evolve through natural selection in this way, the first person in human history to utter a laugh would need to gain some advantage from it. Imagine you are that person: what kind of reception would you expect from your group? Puzzlement, maybe, but not a chorus of laughter in return. So, the first person to laugh would not benefit from the advantage that Dunbar believes drove the evolution of laughter. Thus any advantage to be gained from social cohesion through laughter cannot have been the original reason it evolved. It seems a lot easier to imagine laughter evolving gradually as a play vocalisation and this allowing human groups to expand, than the other way around as Dunbar suggests. In other words, laughter may have been the glue that allowed human social groups to continue to cohere as they grew in size, but can laughter have evolved primarily for this purpose? It seems very unlikely.

To recap: laughter is a play vocalisation that is characteristically human but fundamentally mammalian. Rats laugh, though at a frequency well above the range of human hearing. What the whole troupe of laughing mammals has in common is that we are all social animals that learn how to negotiate life in groups through play. The cohesive effect of laughter may have enabled our ancestors to live harmoniously in especially large groups. One-on-one, there is another kind of social signal of affiliation: the smile. Could smiles be the ancient forebears of laughter?

Chapter Five

Smile and Wave

'Smile first thing in the morning.
Get it over with.'
W.C. Fields

A smile, like a laugh, is an emotional signal recognised across boundaries of language and culture. We associate both with humour, but a smile is much more intimate. A laugh can be heard by anyone close by but a smile is beamed face-on at a target. It is intended and read as a personal message, but what is it saying?

Three bodies, belonging to a Frenchman, a German, and a hillbilly are delivered to the morgue and the coroner notices to her surprise that all the corpses have smiles fixed on their faces. She consults the three death certificates to find out how each man died. The Frenchman's certificate says, 'Cause of death: La petite mort.' Odd, thinks the coroner. 'Little death?' She checks the body. It's majorly dead. 'Ah,' it then dawns on her, 'La petite mort is what the French call an orgasm. That's why this man was smiling when he died.' Next she looks at the German's death certificate. It says, 'Cause of death: heart attack when he saw England knocked out of the Football World Cup.' 'Schadenfreude,' she thinks to herself. Finally she checks the hillbilly's certificate. It says: 'Killed by lightning.' The coroner turns to her assistant for an explanation. 'Why was he smiling?' 'He thought he was having his photograph taken.'

Every smile is the result of the contraction of a pair of facial muscles called the *zygomaticus major*, but the operation of a dozen other facial muscles creates subtle differences of expression between types of smile.[1] With their accustomed fondness for thinking in threes, psychologists recognise three kinds of smile, just like in the joke (strange, that). The Frenchman's smile was of the reward type, signalling his pleasure to his lover. The German was experiencing joy at the discomfort of others (*schadenfreude*) and his was a smile of dominance. The hillbilly thought that he was being photographed and flashed a smile of affiliation.

There are certainly more kinds of smile than just the three, as people may smile shyly, or when they are embarrassed or sad, for example. But, leaving that aside, how do we even know that the reward, dominance and affiliation smiles are genuinely different from one another? How easily can people distinguish between them? Testing this in an independent way is tricky. If you just show people a lot of photographs or videos of people smiling, the results could easily be influenced by the experimenters' own preconceptions of what different smiles look like. Also, the faces in the images will differ from each other in all the ways that people do, and this could influence how people perceive their smiles. Some psychologists (there were six of them, since you ask) overcame these problems in a rather ingenious way.[2]

Computer software that generates lifelike animations of human faces was programmed to display thousands of smiling faces with random variation mimicking the contraction of the minor muscles that produce variation of expression. Volunteers viewed the random faces and were asked to rate each one according to how strongly it suggested a reward, an affiliative or a dominant smile. This procedure selected a subset of the random faces to represent each of the three smile types. The three facial expressions selected by the volunteers differed from one another in recognisable ways. Reward smiles were symmetrical and were accompanied by raising of both eyebrows. Faces with smiles that were identified as affiliative showed lips pressed together. Dominance smiles were asymmetric, with a wrinkled nose and a raised upper lip.

A mixture of these images was then shown to a separate pool of volunteers who were asked to rate to what extent the person depicted was feeling positive, feeling a social connection with someone, or feeling superior or dominant. This group of volunteers, independently of the first group, was able to distinguish between the three types of smile and ascribe the appropriate feelings to each face. Both reward and affiliative smiles were interpreted as showing someone with positive feelings, but reward smiles showed less social connection than affiliative ones. Dominance smiles stood out clearly from the other two types.

Like so many psychological experiments, the one just described may seem to merely prove the obvious — that smiles are subtle social signals that can convey a range of meanings. But we should resist the temptation to believe our preconceptions when evidence may contradict them. We are not, after all, independent witnesses to our own motives and behaviour. Take the reward smile, for example. Is this really a social signal, or is a smile just the natural reaction to pleasure? Think about this for a minute. Perhaps, like me, you'd rather not think you are signalling to anyone when you smile with pleasure. Now consider some evidence.

People participating in sports events smile with pleasure when they score, right? Actually, studies that film exactly what players' faces are showing at the moment of triumph reveal that they very rarely smile.[3] The smile comes afterwards, when the player turns to the crowd or to the team. So, for example, a study at a bowling alley filmed players' faces at the moment they scored a strike, knocking down all ten pins with a single ball. When facing the pins, there was no smile, but when they pivoted to face the other players the smile appeared. The same is true of Olympic medallists, fans watching a football match and others. When interviewed, the subjects all reported feeling happy throughout, but they only showed it with a smile when interacting with other people. A smile is indeed a signal of how you are feeling and not merely a reflexive response to pleasure.

What, then, is the relationship between smiling and humour and can we trace the evolutionary history of the link between the two? Here are some questions we shall try to answer. When was the first smile? What did it mean? Did the three different smiles —

reward, affiliation and dominance — evolve one after the other and if so, in what order? When and how did smiling become associated with humour? Why can a smile mean so many different things?

A psychiatrist who claims that he can tell the frequency with which a man has sex just by looking at the smile on his face goes on a TV chat show to promote his new book. The chat-show host asks for a volunteer from the studio audience and a guy with the widest of grins comes forward. The psychiatrist studies the man's smile and says:

'Twice a day.'

'Nope.'

'Daily.'

'No.'

'Twice a week.'

'Nope.'

'Weekends.'

'Nope.'

'Monthly.'

'Nope.'

'OK, OK. I give up.'

'Once a year.'

The shrink explodes: 'Then why the hell do you have such a big grin on your face, man?!'

'Tonight's the night.'

A Chinese proverb says that food can be enjoyed three times: in anticipation, in participation and in memory. The shrink forgot that the same goes for sex.

One reason a smile might have so many meanings is that it is anciently evolved. The longer it has been in our behavioural repertoire, the more opportunities there would have been for evolution to find new uses for the signal. If this hypothesis was correct, we'd expect to see smiles like ours in the behavioural repertoires of chimps and other great apes. They do have a play face that is functionally equivalent to a smile, but this relaxed open-mouth display is not formed by the *zygomaticus major* muscles that make us smile. So, just as human and chimp laughter achieve the same function by different means, so too do human and chimp smiles. What this means is that the human smile as we know it probably evolved after our lineage diverged from the last common ancestor with chimps about 6.5 million years ago.

Considering how many big changes happened in human evolution in the last 6.5 million years — walking upright, nakedness, a change in diet, an increase in brain size and intelligence, the evolution of speech and humour — there has been plenty of time for smiling to diversify too, but there is another reason a smile can mean so many different things. A smile can be subtly moderated by various muscles in the face. Indeed, a good actor can convey conflicting and changing emotions with the face alone, without uttering a word. Thus, a smile is an inherently changeable expression to which various meanings can easily become attached.

Charles Darwin suggested how emotional expressions could evolve, starting from a gesture that has a specific function such as aggression that others can see is a warning. In social encounters between animals, bared teeth are an obvious threat. If bared teeth are a clear sign of aggression, then covering the teeth with a lip press can be read as the opposite — and this may be how the affiliative smile evolved. Once affiliative smiles are established, modifications could arise conveying other meanings. Which came next, reward or dominance? We don't know, but given that affiliative and reward smiles are perceived as similar and that evolution proceeds in small steps, it is natural to assume that the reward smile came next. As for the dominance smile, only after a social or family group has been cemented with affiliation is it possible for its members to contest for dominance. So, it seems logical to propose that dominance smiles evolved last, though it is also quite possible that it was contemporaneous.

What is the evolutionary relationship between smiling and laughter? A number of authors have proposed that smiling is essentially equivalent to laughing, but with the sound turned off and the emotional amplifier turned down.[4] A weak joke might make you smile, but a good joke will make you laugh. This might suggest that smiling, the weaker signal, evolved first and that laughter followed. However, the comparative evidence suggests otherwise. As we have seen, play vocalisations such as laughter are general in social mammals and therefore must have evolved in our ancestors long before primates, let alone great apes, appeared. Although dogs and apes have play faces, they do not have the flat face of our own species that lends itself so well to the diversity and nuance that can be conveyed in a smile.

We don't actually know whether smiling or laughter evolved first, but they appear in a very definite order during child development. Babies begin to smile at about four weeks, but do not begin to laugh until four months. Smiling between mother and infant is essential to creating a bond between them. The maternal bond is so obviously important to the survival of an infant that this might have been the original evolutionary advantage of smiling. But if that was the case, how did it get started? Just as with the first laugh, the first reward smile from a baby would have been met with blank incomprehension unless it was in some way related to an existing signal that the mother could recognise. If you have ever cared for a newborn and had the conversation with your partner about whether what looks like baby's first smile is actually wind, you will understand the problem.

As for affiliation smiles, these appear at the age of about two, but dominance smiles do not occur till adolescence and by that age culture plays a big part in how humour develops.[5] You will have noticed that there has been rather little mention of humour in this discussion of smiling. This is because smiles fulfil so many social functions besides advertising amusement that it is not even clear whether there is such a thing as a pure smile of amusement. Could a smile of amusement just be what we call a reward smile when humour pleases us? If smiles are ambiguous, intimate and subtle, the same cannot be said of laughter. Quite the reverse, in fact. Laughter is an unambiguous broadcast of mirth to anyone in earshot. Why the difference? There is good evidence that the evolution of humorous laughter depends on one dominant function: sex.

Chapter Six

Laughter and Sex

We primates have been laughing at life and at each other for a very long time. It seems likely that it all began with a tickle in play and this legacy is still a part of our everyday lives. Laughter is heard 30 times more frequently in company than when you are on your own. In conversation, the speaker laughs more than the listener, indicating that it is a normal part of verbal communication. If you are chuckling to yourself, you probably have virtual company in your head or on the page or screen. Hearing laughter induces laughter in others. All of this play vocalisation can happen without anyone even saying, 'Have you heard the one about the three-legged chicken?' When you do hear that joke and laugh, you are responding to something different: incongruity. But why is resolving incongruity funny?

This question turns out to be easier to answer once we recognise what incongruity actually is. An incongruity is a mismatch between sensory data and expectation. For example:

> *A little girl goes into her local library to take out a book called* Advice for Young Mothers. *'Why do you want a book like that?' asks the horrified librarian.*
>
> *'Because I collect moths,' replies the little girl.*

The mental faculty that detects incongruity is an error-detecting mechanism. This hypothesis has itself been gradually evolving over the centuries, through the humour theories of Aristotle, Hobbes and Kant amongst others, coming to fruition in our own century when three cognitive scientists called Hurley, Dennett and Adams decided not to walk into a bar, but to write a book instead. In *Inside Jokes: using humour to reverse-engineer the mind,* they use humour to look inside our heads. It has to be said that fMRI works better, though as we have seen, using the two together works best of all. The insight that Hurley & friends brought to the party was essentially that errors are bad and so spotting and debugging them is good. There! I've saved you 300 pages of dense reading.

To anyone who has ever been on the wrong end of a piece of failing computer software, it will be self-evident that having an error-detecting mechanism is a good thing. Just think of the mess that predictive text can make of a message written on a smartphone, such as this classic:

Great news — Grandma is homosexual!

Okay?

Homo hot lips
Hot tulips
I am getting fisted
Frustrated
Grandma is h o m e from h o s p i t a l

When the phone's software fails, our own error correction circuits are tickled into action. The humorous response to incongruity is how the mind identifies cognitive bugs. It is obviously a good thing to have an error-detecting algorithm running inside your head, so it's not hard to see why natural selection would have endowed us with this. Just to make sure that we do what is good for us, Hurley *et al.* contend that 'the pleasure of mirth is an emotional reward for success in the specific task of data-integrity checking.'[1] Or in other words, evolution says, 'Do your homework, it'll be fun!' I'd like to offer you the alternative hypothesis that when evolution attached humour as a trigger to laughter, fun came as part of the set. It was there already. Batteries included, ready to play.

However, there is something not completely convincing about the debugging hypothesis itself and it is this. If laughter just signals debugging, why do we laugh only at *frivolous* errors?[2] Surely, if the pleasure of laughter is a reward for successful debugging, then the more serious the error, the bigger should be the laugh. But it's not actually like that. Make a more serious mistake, like forgetting your partner's birthday, and neither of you will laugh at the error.

It is as though evolution has designed us a crucial piece of software that can spot errors and save lives, and then rewarded us for just playing around with it. So, could there be another, better explanation for why we find the resolution of incongruity funny only when, as Darwin observed, it is not momentously important? The answer is yes, and it's to do with sex.

The sex hypothesis is that humour is a public display that influences courtship. It was first proposed by psychologist Geoffrey Miller,[3] based on an insight that yet again we owe to Charles Darwin. In 1871, the year before Darwin published his book *The Expression of the Emotions in Man and Animals*, he published another ground-breaking volume called *The Descent of Man, and Selection in Relation to Sex*.[4] It's a curious work that is really two books in one. The first part is all about establishing that humans belong to the animal kingdom and that we're not specially created, while the second half of the book is a comprehensive survey of all the cases Darwin could find where males and females of the same species differ from each other in secondary sexual characteristics. These are the characteristics, other than reproductive organs themselves, that distinguish the sexes. Boobs and beards in our own species for example.[5] Why, Darwin wanted to know, should the sexes so often differ like this?

Darwin's answer was that the evolution of secondary sexual characteristics can be explained by the different requirements of acquiring a mate. Natural selection will favour any characteristic that helps in courtship. Darwin called this kind of natural selection 'sexual selection'. Geoffrey Miller reasoned that many cultural features of human behaviour, like music, art and humour, may have arisen through sexual selection because they help attract mates. One potential difficulty with Miller's hypothesis is that artistic ability does not necessarily differ between men and women, but we shall return to this point.

The reason that we expect sexual selection to produce differences between males and females is the fundamental fact that sperm are cheap and plentiful compared to eggs. A man produces new sperm daily, but a woman is limited to the number of egg cells she was born with. This difference, and more especially the greater biological investment required for a female to gestate her offspring than for a male to sire one, has profound evolutionary consequences. It is also a fertile source of jokes:

> *My wife only has sex with me for a purpose. Last night she used me to time an egg. (Rodney Dangerfield)*
>
> *The morning after the wedding night, the bride says, 'You know, you're a really lousy lover.' The bridegroom replies, 'How can you tell after only 30 seconds?'*

While, from a female's perspective, reproduction looks very different:

> *To simulate the birth experience, take one car jack, insert into rectum, pump to maximum height, and replace with a jack-hammer. And that would be a good birth. (Kathy Lette)*

Females tend to be choosier than males in their selection of mates because of their greater biological investment in offspring. Dorothy Parker had a keen sense of this imbalance. She said after an abortion 'It serves me right for putting all my eggs in one bastard.' Parker named her canary 'Onan' because, like the character of that name in the Old Testament, 'He spills his seed on the ground.'[6] Only a male can afford to be so profligate.

Why do males produce millions of sperm? Because they won't stop to ask the way.

Of course, some religions do not agree about the superfluity of sperm, as satirised by Monty Python in their song 'Every Sperm Is Sacred'.[7]

Sexual selection operates in two distinct ways: through competition between members of the same sex and through mate choice between the sexes. Competition between males for access to females is very common and explains, for example, the large tusks of bull elephants and the antlers of male deer used in battles between stags during the rut. Female-female competition can occur too of course:

Joan Rivers: Besides your husband, who's the best man you've ever been in bed with?
Joan Collins: Your husband.
Joan Rivers: Funny, he didn't say the same about you.

Some say that female fashion is based upon competition between women. Groucho Marx once quipped:

> *'If women dressed for men, the stores wouldn't sell much — just an occasional sun visor.'*[8]

Sexual selection is often driven by the choosiness of females for mates and it can lead to the evolution of extremes in males. The peacock's tail is a good example that was originally discussed by Darwin himself. Peahens are brown, dowdy-looking birds that bear little resemblance to their richly ornamented husbands. It has been shown through experiments in a free-ranging population at Whipsnade Zoo in England that peahens prefer to mate with males who have more elaborate displays and that peahens lay more eggs for their preferred mates than for other males.[9] Other studies of peafowl in Japan and Canada failed to find the same female preference,[10] so it is possible that peacocks are now stuck with a trait that females no longer use to choose a mate, although peahens won't look at a male who can't at least shake a tail feather. Equally, it may just be another case of a general phenomenon that plagues studies of animal behaviour:

> Under carefully controlled experimental
> circumstances, an animal will behave as it
> damned well pleases.

We don't know exactly how the female preference for showy males got started in peafowl. Perhaps, in the beginning, the healthiest males were naturally bright-eyed and bushy-tailed and females recognised these first modest signs of quality and responded to them. However the female preference started, once it did so a runaway process of evolution was begun in which females who chose showy males for mates would have male offspring who inherited the secret of success with the ladies. That would then perpetuate the advantage of the female preference and lead to continuing sexual selection for more and more showy tails in peacocks.

It is important to evolution through mate choice that the cue that females use to pick the best males cannot be faked by inferior males. The cue needs to be correlated with quality as a mate. Only well-fed, disease-free peacocks can produce a really big and sexy train. It is significant to the evolution of indicators like a peacock's train that they are costly to the male because this guarantees that the signal is genuine. Otherwise, cheats could prosper in mating and sexual selection for showy tails would cease to work. Imagine if peacocks could acquire a clip-on train, what havoc this would create for female choice.

Ornamented males with plain-looking females is a common pattern among birds, suggesting that mate choice by females is an important form of sexual selection in these species. Mate choice can be exercised in the other direction as well, by males for females:

A businessman has decided that he must finally marry and he needs to choose between three women whom he has been dating. His golf buddy suggests that he set them all a test to find which one would be the most suitable. 'Give each of the ladies £5,000 and see what they do with it. Then decide.' The businessman likes this idea, so that is what he does.

A month later, the two men meet again and over a round of golf the businessman tells his friend what happened. 'Jane took the £5,000 to Bond Street, bought herself an engagement ring and proposed to me,' he said.

'Wow, she sounds keen! What did the others do?'

'Elizabeth is in finance. She invested the money in the stock market and this week she sold her shares at a 50 per cent profit. She kept her profit and gave me back my £5,000.'

'What a gal! She's the one, right?'

'Fiona went out and bought herself some sexy lingerie, booked us into a luxury spa hotel in the Cotswolds, told me how much she loved me, and we had the most amazing weekend.'

'Man, these women are all keepers. So who are you gonna choose?'

'The one with the biggest tits, of course.'

So how, you will by now be asking, can sexual selection favour humour? The answer is that humour could be a marker for intelligence that is favoured by either sex when choosing a mate. Here is how it might work. We start with the observation that, judging by its results, human evolution clearly favoured intelligence. From there, it's a small step to the hypothesis that intelligence is a desirable trait in a mate — why would you want your children to be dumb members of a smart species? Surveys of mate preference show that women do indeed prioritise intelligence in their choice of long-term partner.[11]

This is no doubt why the Irish playwright George Bernard Shaw (1856–1950) was once propositioned by a famous actress with the line, 'Think of the child with your brains and my beauty,' to which he replied, 'But what if he were to have your brains and my beauty?' Shaw's reply illustrates the next step in the argument, which is that wits and wit are correlated.

Clever people make good jokes. And there you have it. A person's capacity for humour is their peacock train — the un-fakeable evidence to any mate who will listen that you are one sexy beast who will produce clever children.

'Hmm,' you should be thinking to yourself at this point. 'Here is an author cracking jokes about how clever and sexy people who crack jokes are. Could this not be a tad self-serving?' And you would be right to ask that, so we must look at the evidence. But before we test out the idea that humour has evolved as a public display of intelligence influencing courtship, let's caution that people are not peafowl. In peafowl, it's the male (the peacock) who has the fancy display, but in humans both sexes participate in humour. For that reason, our hypothesis is that sexual selection operates in both directions in humans, arising from mate choice by both sexes. Whether it is equally important to both is another question that we will consider in due course.

First, we must determine whether intelligence and humour ability are inherited to any degree. If either is not heritable, they cannot evolve, and the hypothesis must be rejected right there. Next, we need to see whether humour ability is correlated with intelligence, because if it is not, the sexual selection hypothesis is holed beneath the water line. Then, we need to know whether humour ability affects mate choice and how it does so. If people see jokers as fun and good for a fling, but not reliable enough to be co-parents, for example, our hypothesis would fail. If the hypothesis passes all those tests, we must then investigate how sexual selection on humour could have started. It's a long, long way from a tickle more than 9 million years ago to the craic of verbal humour in the pub.

The heritability of any trait such as intelligence or humour ability can be estimated by comparing twins.

Man: My wife is a twin.
Friend: Really? That could be embarrassing. How do you tell them apart?
Man: Her brother has a beard.

Some twins are identical (monozygotic) because they originate from the division of a single fertilised egg in their mother's womb, while others are non-identical (dizygotic) because they developed from two different fertilised eggs. Dizygotic twins are genetically no more alike than ordinary siblings. The logic of the test for heritability is that if genetics influences intelligence (IQ), say, then identical twins should have more similar IQs than non-identical twins. When raised together, both kinds of twin share environmental influences with their sibling, but because monozygotic twins are genetically identical as well, they should be even more alike if genes have influence.

The relative influence of nature (genes) and nurture (environment) on IQ has been controversial for decades, but luckily, we need not get bogged down in that argument because the evidence of twin studies clearly shows that both genes and environment influence all important psychological traits, including IQ.[12] There only has to be a small significant effect of genes on intelligence for selection to be able to alter it over the generations. The IQ controversy has been about *how much* influence genes have — but this is not a useful question because a person's actual intelligence is determined by an interaction between their genes and their environment.[13] For example, genes influence educational attainment, but good education can compensate for a poor inheritance.

An interesting question is why, if cognitive ability (intelligence) is so important to fitness, is there any genetic variation in this trait still remaining when natural selection must have been favouring the brainiest for millions of years. Surely all the good genes should have replaced all the deleterious ones by now and we should all be as clever as Albert Einstein and as witty as Dorothy Parker? The reason that this has not happened is most probably because about a thousand different genes are known to influence cognitive ability and they are not all pulling in the same direction at any one time. Trade-offs between beneficial and deleterious effects of genes are common and they preserve the genetic variation upon which natural selection may act.[14] So, we'll run out of jokes before we run out of genes that influence the intellect.

Is a person's ability to be funny in any degree inherited? Twin studies have asked people how often they use and enjoy humour, but have not objectively evaluated the subjects' ability to be funny. Two studies conducted in North America failed to find a genetic component to self-declared humour appreciation, but studies in the UK and Australia found that between 30–50 per cent of the variation in how much people said that they enjoyed humour was genetically influenced.[15] The explanation for this difference between nations probably lies in the kind of humour that people enjoy in different cultures, because this would have influenced how they answered questions in the test. Pretty much every other twins test on personality differences has found some influence of genetics, so it's probably safe to assume that sense of humour is no different.

The sexual selection hypothesis has now hurdled the first two fences, though only just scraping over the second one representing the heritability of humour ability. We could do with some more direct evidence on that one. Next up is the correlation between humour ability and intelligence. We are going to sail right over this challenge with the help of two studies of university students taking psychology courses.[16] The ability of each student to produce humour was tested objectively and compared with their scores in tests of cognitive ability. In both studies, cleverer students were funnier and this applied to both sexes. It's unfortunate that most psychological studies are conducted with students because they tend to be from WEIRD (Western, Educated, Industrialised, Rich, and Democratic) societies, and are not a random sample of humanity.[17] However, until there is evidence to the contrary, we will accept what these studies tell us, which is that wit is a genuine indicator of wits.

Dead ahead now is the question of whether there is a mating advantage to being funny. One of the student studies mentioned looked at this with a questionnaire on frequency of intercourse and number of heterosexual sex partners, finding the predicted link between mating success and humour ability.[18] This supports our hypothesis, but it's only a correlation. With correlations, there is always the possibility that the real reason that jokers get laid more is something else that we have not taken into account, for example that they happen to be taller or better looking.

Is there more direct evidence of a causative link between humour ability and mate choice by the opposite sex? A study in France tested this connection with an experiment conducted in a number of bars (where else?). Conversations were staged between three young men in which one of them told jokes in a loud voice that could be overheard by a nearby young woman.[19] After the conversation, two of the men left the bar and the remaining man went up to the young woman and asked for her phone number. Joke-tellers were twice as successful in obtaining phone numbers as non-joke tellers. While this experiment does directly support the idea that women prefer funny men, its success was surprising for two reasons. First, because the jokes themselves were so lame. It is only out of necessity I reproduce one of them here:

> *Two friends are talking. 'Say, buddy, could you lend me 100 Euros?'*
>
> *'Well, you know I only have 60 on me.'*
>
> *'Ok, give me what you've got and you'll only owe me 40.'*

And second, because the man telling the joke always began by saying to his friends, 'I've got some good jokes to tell you,' so he was obviously not demonstrating his native wit, but his memory. Perhaps the jokes sounded better in French.

Finally, when it comes to the question of whether humour influences mate choice, we cannot ignore the fact that a Good Sense of Humour (GSOH) is a common characteristic that is both sought and offered in advertisements for partners, or at least it was before smartphones and dating apps like Tinder came along.

A guy complained to a dating agency that they had not matched him with any compatible dates. 'Haven't you got someone who doesn't care what I look like, isn't bothered that I have no sense of humour, and has a lovely big pair of boobs?'

The woman running the agency checked her database and replied, 'Actually we do have one. But it's you.'

Most people think they have an above-average sense of humour. Since it is statistically impossible for everyone to be above average, it just goes to show how highly we rate being funny. Nobody advertises that they want a good sense of smell or an above average sense of taste in a potential mate. Women advertising for a male partner ask for GSOH more often than men do when looking for a female partner. Men also value a sense of humour, but apparently not as much. What they really want is a woman who will find them funny.[20] Because, as Virginia Woolf sardonically observed:

> Women have served all these centuries as looking glasses possessing the magic and delicious power of reflecting the figure of man at twice its natural size. Without that power, the Earth would probably still be swamp and jungle.'[21]

It's important to consider whether some of the apparent difference between how men and women value GSOH in a mate could be cultural rather than genetic. An online survey of mate preferences of more than 200,000 people in 53 nations found that sex differences were fairly consistent across countries. From 23 traits describing a partner, the top three for men were intelligence, good looks and humour. The top three for women were humour, intelligence and honesty.[22] It is noteworthy that intelligence and humour are valued by both sexes in choosing a mate and that the difference between the sexes is only one of small degree.

We also have to recognise that national differences are not the only source of cultural influences on humour. Most nations are patriarchies in which men are in control. There is a preconceived idea that men are funnier than women.[23] It is generally men who believe this but they don't experience humour between women, who actually laugh more in conversation than men, especially when in all-female company.[24]

Two members of the Women's Land Army are digging in a carrot field during World War II. One picks up a giant carrot and turns to the other. "Ere, Gert, this carrot reminds me of my old man!'

'Corr,' says the other, 'is he really that big?'

'Nah, he's that dirty.'

The rising number of successful female comedians now holding their own in a business still heavily dominated by men shows, if you ever doubted it, that female humour was previously hidden, not missing.

The sexual selection hypothesis has successfully jumped all the tests that we have put in its path and so it seems that mate choice could indeed explain why we use humour to draw attention to our sexy cognitive prowess. Sexual selection for humour ability can in theory operate for both male and female mate choice because the situation in humans is different from that in peafowl.[25] A peacock's only contribution to his offspring is his genes. He does not participate in raising his offspring as males of other bird species like the blackbird do, or indeed as is the norm in human societies. If all you have to offer a mate is a resource as abundant as sperm, you need to really shake your money-maker to get attention. But, if the sperm come with a care package for mother and offspring, this is of more value to a female and consequently males can exercise mate choice too. Men are able to exercise mate choice because they do have more to offer women than just their genes.

> *'There are a number of mechanical devices which increase sexual arousal, particularly in women. Chief among these is the Mercedes-Benz 380SL convertible.' (Steve Martin)*

There is another crucial difference between the peacock's train and the human brain. The peacock is advertising something that belongs to and is of use only to a male, but humour is advertising cognitive ability, which is manifest in females as well as males and is of value to both. With a thousand or more genes involved in cognitive ability, these cannot be confined to that tiny part of the genome on the Y chromosome that is the exclusive property of males. So, genes affecting cognitive ability are necessarily present in, and are expressed by, both sexes. The actress who propositioned George Bernard Shaw was absolutely right. If she had borne his child, whether a boy or a girl, it would have inherited a share of his intelligence and her beauty because both are determined by many genes. GBS may have been a wit, but he had some half-witted ideas about genetics.

Peacocks may be only an imperfect model of sexual selection as it applies to humour, but there is a feature common to both that we have not discussed yet.

Why didn't the peacock cross the railroad track?
Because he didn't want to catch his train.

Humour, like a peacock's train, can be a handicap if you get it wrong. This joke ought to be funny, but it's a groaner, isn't it? I'm not sure why, perhaps because there is insufficient incongruity between the set-up and the punchline, or because it's too complicated. You can see that it's meant to be funny, but it's a flop. And there is the hazard of humour. Failure is as public as success. Remember that for sexual selection to work, the production of the indicator must be costly to the producer or cheats will defeat the discriminating value of the indicator. The cost of showing off your cognitive ability with humour is that you expose yourself to ridicule as well as to approval. The cost of trying to make people laugh is that you yourself might end up becoming the joke and feeling stupid.[26] This hazard is inherent in humour and only the genuinely witty can avoid it. Fakes will out. People are adept at telling the difference between genuine and forced laughter, even across a language barrier.[27]

Let's summarise the story so far before we move on. All young mammals play and emit play vocalisations that may have originated as an 'all clear' signal. This seems to have been the evolutionary origin of laughter, and may also explain why laughter is contagious, since play requires all playmates to signal their harmless intent to each other.[28] Humour later became an additional trigger for laughter, inheriting the properties of pleasure, safety, spontaneity and contagiousness that are associated with the original play vocalisation. Humour is generated by the resolution of incongruity between expectation and sensory data. This is a kind of de-bugging mechanism, but that cannot be the main function of humour since it is only triggered by incongruities that are non-threatening. An alternative hypothesis that is supported by a wide range of data is that humour is a public display of cognitive ability favoured by sexual selection through mate choice.

So far so good for the sexual selection hypothesis, but what about another possible benefit of humour, the idea that 'Laughter is the best medicine.' Taken literally, this idea is not true. A survey of the health of 500 amateur stand-up comics, who must surely be exposed to more laughter than anyone else, found that they were actually in worse health than people of similar age and gender.[29] In fact, there is good evidence, taken from comparing the lives of comedians and other actors, that comics die younger. Worse still, funnier comics die youngest of all, though why is not yet clear.[30] The British comedian Tommy Cooper actually died in the middle of his show at the age of 63. None of these studies were clinical grade, double-blind trials of the kind used to test the efficacy of actual medicines, but they suggest that stand-up comedy is not the healthiest of professions, even by the low standard of performers more generally. Comedians are laughter-makers. No one to my knowledge has yet tested whether audience members live longer if they laugh more.

The phrase 'Laughter is the best medicine' has an evolutionary history of its own, having started in Proverbs 17 of the Bible, which in the King James version reads 'A merry heart doeth good like a medicine.' This more modest claim is supported by the scientific evidence, though with some qualification. Although uncontrollable laughter has physiological effects and can even render the laugher temporarily helpless, there is no evidence of physical benefits from laughter as exercise.[31]

In contrast, the beneficial effects of laughter on feelings of well-being and mental state are well supported.[32] Not only does physical laughter cheer people up by generating endorphins, but laughter also raises the pain threshold.[33] The cognitive effort involved in cracking jokes in stressful circumstances can help not only during the stressful situation itself, but also with coping with the memory of it afterwards.[34]

Is it possible that the beneficial effects of laughter on mental health could have contributed to its evolution in the first place? That does seem possible, though it would be additional, rather than an alternative to the role of sexual selection. Play and choice of mate between them account for so many features of laughter that health benefits alone cannot explain on their own. For example, in long-ago ancestors who had not yet evolved a sense of humour or attached this to laughter, what health benefit would detecting trivial incongruities provide? On the other hand, once humour and laughter had evolved with their associated hormonally-induced feel-good reaction, it is easy to see how the resulting beneficial effects on mental health might add to the advantage of humour. It is typical of evolution to tack one benefit onto another, and this may be what happened with health effects, although it is equally possible that these are merely fortuitous by-products.

In the story so far, we have given the biology of laughter a thorough examination and uncovered its hidden evolutionary history.

What do you call someone who does not believe in evolution?

A primate change denier.

The most important step is to realise that *'Why did evolution make us laugh?'* is a meaningful scientific question in the first place. Once you realise that it is, it's only a short step further to seeing that laughter is a play vocalisation with a deep evolutionary origin. Humour became a trigger for laughter much more recently, but it built on another pre-existing function of the brain: error detection. Curiously, we only laugh at trivial errors, or incongruities, because laughter is not about survival but seduction, not about defence but display. Sexual selection is the key: humour is a demonstration that you have the wits to woo.

Chapter Seven

Jokes and Culture

I want to die peacefully in my sleep like my grandfather, not screaming in terror like his passengers.

This was the favourite joke in Scotland according to a huge online survey called LaughLab that was run in 2001 to discover the world's funniest joke.[1] LaughLab received 40,000 jokes and 350,000 ratings from people in 70 countries. The joke rated top by people in England was a different one:

Two weasels are sitting on a bar stool. One starts to insult the other one. He screams, 'I slept with your mother!' The bar gets quiet as everyone listens to see what the other weasel will do. The first weasel again yells, 'I SLEPT WITH YOUR MOTHER!' The other says, 'Go home, Dad, you're drunk.'

Americans, favouring dark humour and sports, liked this one the best:

A man and a friend are playing golf one day at their local golf course. One of the guys is about to chip onto the green when he sees a long funeral procession on the road next to the course. He stops in mid-swing, takes off his golf cap, closes his eyes, and bows down in prayer.

His friend says: 'Wow, that is the most thoughtful and touching thing I have ever seen. You truly are a kind man.'

The man then replies: 'Yeah, well, we were married 35 years.'

The Canadians were tickled pinkest by a joke at the expense of their superpower neighbour:

When NASA first started sending up astronauts, they quickly discovered that ballpoint pens would not work in zero gravity. To combat the problem, NASA scientists spent a decade and $12 billion to develop a pen that writes in zero gravity, upside down, underwater, on almost any surface including glass, and at temperatures ranging from below freezing to 300° C. The Russians used a pencil.

Of all the nations participating in LaughLab, it was the Germans, with their poor reputation for humour, rather than the English, who pride themselves for it, who were the most thoroughly amused. The Germans' favourite was:

A general noticed one of his soldiers behaving oddly. The soldier would pick up any piece of paper he found, frown, and say: 'That's not it,' and put it down again. This went on for some time, until the general arranged to have the soldier psychologically tested. The psychologist concluded that the soldier was deranged, and wrote out his discharge from the army. The soldier picked it up, smiled, and said: 'That's it.'

After a lot of jokes, a lot of fun and some number-crunching, psychologist Richard Wiseman who devised the LaughLab came to the conclusion that the world's funniest joke does not actually exist.[2] Humour varies and has a strong cultural component that differs from one nation to another. In the Olympics of humour, every nation plays by its own rules. This perhaps obvious conclusion is made more interesting by the fact that we now know that the underlying brain mechanism triggered by humour is the same, even though the jokes are different. The differences are cultural.

Humour is so integral to human communication and social intercourse that for millennia its core of incongruity has lain hidden beneath an overgrowth of culture. Although great thinkers since Aristotle at least have occasionally glimpsed the truth, there were always rival explanations such as the superiority theory that just could not be dismissed. But now that we have uncovered the essence of humour and established its biological and evolutionary foundations, we can more clearly identify some of the cultural embellishments that have grown up around it for what they are: social phenomena.

The aim of this chapter is to find out how incongruity resolution expresses itself in different cultures and whether there are some rules governing this. We are going to do this by looking at the things that laughter does in society, or in other words its social functions. Social psychologists suggest that laughter, like a smile, does three principal things: it, one, provides a reward that reinforces social ties; two, fosters affiliation among group members, and three, signals the superiority of one group over another.[3] In addition, we must surely add a fourth social function of laughter: subversion — humour as the weapon of the underdog. Let's consider the four functions and seek them out in the jokes of different cultures.

The first thing to note is that the word 'function'

means very different things in biology and in social psychology. In biology, if we say for example that the function of the kidneys is to remove waste products from the blood, this is a statement about the physiological role kidneys play in how your body works. Physiological functions are easily determined by their consequences for health and disease. If your kidneys malfunction, you will need a transplant, dialysis, or an undertaker. Natural selection, the motor of evolution, runs on the power of individual advantage, so it's easy to see why kidneys evolved. Even slugs have them, but I digress.

Social behaviours like laughter also evolve through natural selection, though it is often more difficult to work out exactly how. It is only very recently that we have fully understood how humorous laughter could have evolved through sexual selection. The reason that the evolution of social behaviour is harder to figure out is that it does not operate like physiological functions in one body, but instead across two or more minds. This introduces an extra level of complexity because the advantage to the individual depends on the response of others.

Of the four social functions of laughter, reward is the easiest to understand because of its evolutionary origin as a play vocalisation (Chapter 4). Play is important to the social development of animals, with clear benefits to individuals in later life. Play is how we learn to interact safely with other members of our species on whom we depend for survival and reproduction.

The reward function of laughter is the basis of its second social function, creating a sense of affiliation in a group. We can trace this effect directly back to the origins of humour in play, where laughter is the glue necessary for social coherence in the group. It's easy to see how affiliation is of benefit to the individual in social situations such as in the Glasgow pub displaying this notice:

Everyone brings happiness to this pub.
Some when they arrive
Some when they leave.[4]

Social laughter generates endorphins that create feelings of attachment and well-being, and the consequences of this can be measured. In a study of nearly 100 psychology majors, the lab rats of humour psychology, students who were initially strangers to each other, were matched in pairs of the same sex.[5] Half the pairs were given silly tasks to do together, like learning dance steps blindfold, while the other pairs did similar tasks, but in a serious manner. Interviewed afterwards, the pairs who had laughed together over silly tasks felt significantly closer to each other than participants who had shared serious tasks. Furthermore, the effect was strongest in those students who had tested well for a sense of humour.

Another study took a different approach, observing the role that laughter played in the everyday life of 162 students over a period of two weeks.[6] Each participant kept an online record of any social encounters they had that lasted ten minutes or longer. More than 5,500 encounters were recorded over the two weeks, describing in each case whether there was laughter, how positively the subject felt about the interaction and whether it was with someone they had met previously, or someone new. The study found that laughter in a social encounter with one person improved the experience of the next encounter, regardless of whether it was with the same person or with someone new. Subjects felt a greater sense of intimacy with anyone they met if they had laughed previously. The effect did not work in reverse. Subjects who met with a person they felt intimate towards were not more likely to laugh in a subsequent meeting. In other words, it was laughter that created the sense of closeness to others, not the other way about.

The results of studies like these should not really surprise us, but they provide valuable scientific confirmation that laughter spreads positive interactions among people in daily life. More surprising is the finding of a huge study of nearly 5,000 people monitored over two decades that suggests that happiness is contagious and that not just the people *you* meet, but the people *they* meet are also positively influenced by your good humour.[7] The study was conducted in the small town of Framingham in Massachusetts, where inhabitants have for three generations volunteered to participate in a long-term medical study of heart disease. In addition to regular health checks, the Framingham Heart Study recorded friendships, family relationships, neighbours, and co-workers. Everyone in the study regularly answered a questionnaire that included questions designed to detect depression, with options at the happier end of the scale allowing people to report that in the previous week 'I felt hopeful about the future', 'I was happy', or 'I enjoyed life.'

This study was not originally designed to test how people's mood affected that of others in their social network and we do not know whether people who said that they were happy actually laughed with their friends and relations, though it would be odd if they didn't. Nonetheless, what researchers discovered is that happiness was clustered in the Framingham population. Happy people were socially connected to each other. This observation alone might be due to the 'birds of a feather flock together' effect. In other words, happy people gravitate towards other happy people.[8] This would certainly be the simplest explanation, but because the Framingham population was surveyed repeatedly over many years, it was possible to test an alternative hypothesis: that changes in the happiness of one person affected changes in the people they knew. This was indeed what the results suggested.

When a person became happy, friends living within a mile of them were 25 per cent more likely to be happy. A similar effect was seen in next-door neighbours, and similar though smaller effects were seen in spouses living at the same address and siblings living nearby. An effect of one person's change in happiness was detectable not only in their immediate friends, but up to three degrees of separation away: that is in their friends' friends' friends. The further away friends and relations lived from the happy person, or the longer the interval between contacts with them, the smaller were the positive effects.

When the study showing that happiness could ripple through a community was published in 2008, it caused a big stir, and as always happens in science, there was healthy scepticism that something simpler and less remarkable was happening than what was being claimed. Happy people know happy people would not really be news, would it? On the other hand, a ripple of happiness is just what you would expect to find occurring, given what we know about how laughter by one person improves the mood of the people they meet. However, we need to guard against confirmation bias. The mere fact that we get a result that makes sense to us, does not make it true.

I caught my neighbour sprinkling flour over his car. 'What are you doing that for?' I asked.

'Well,' he said, 'it keeps the polar bears away.'

'But, there are no polar bears round here,' I said.

'Exactly!' he replied. 'You see, it works!'

The only way to be sure that one person's happiness is the cause of happiness in their friends, and so on to the third degree is to perform an experiment in which you manipulate the mood of a random sample of people and look for resulting ripples in their social networks, comparing them with an unmanipulated, control group. The Framingham study came near this experimental standard because it did follow changes in happiness over time. However, the initial causes of the changes in happiness were unknown, and not due to deliberate experimental intervention in the lives of the population. Of course, you cannot ethically manipulate people's mood without their consent, but the consequence is that we cannot be sure that the effects seen in this study are genuinely the result of contagious spreading among friends and neighbours. They might equally be the effect of shared experience. Events such as weddings could elevate the mood of a whole social group, for example.

The hallmark of science is reproducibility, but the Framingham study is practically unique, so social scientists have had to look elsewhere for evidence that emotions like happiness are contagious in social networks. The evidence has been found on Facebook. A study by a Facebook data scientist and two colleagues at Cornell University conducted a massive online experiment in which the news feeds of nearly 700,000 users were manipulated by filtering posts to include less positive, less negative or an unchanged emotional content.[9] The results showed that people on the receiving end of the experimental manipulations responded by posting fewer positive or fewer negative messages of their own, in line with what they received from others. Because these changes were responses to experimental manipulation, the study conclusively demonstrated that emotional contagion was operating in Facebook's social network.

Unfortunately, participants in the Facebook study were not specifically informed that their newsfeeds were being manipulated and their responses monitored, so they could neither give informed consent nor opt out. This contravenes the ethical principles normally applied in social science, though it was covered by Facebook's terms of data use at the time. After publication, and when the ethical horse had bolted the stable, the very prestigious scientific journal that published this study issued a polite editorial 'expression of concern'. As long ago as 1946, the poet W.H. Auden anticipated this kind of compromise of academic standards. In a poetic commencement address at Harvard, he admonished graduands with a kind of ten commandments of academic freedom which included the words:

Thou shalt not worship projects nor
Shalt thou or thine bow down before
Administration.
Thou shalt not answer questionnaires
Or quizzes upon World-Affairs,
Nor with compliance
Take any test. Thou shalt not sit
With statisticians nor commit
A social science.[10]

There are ethical ways to commit social science online. An ingenious study avoided the ethical problem of deliberately manipulating users' Facebook feeds by letting the weather do this instead. The study found that people who lived in cities where it was raining at the time they posted on Facebook tended to express more negative emotions and that this increased the chance of friends and friends of friends posting negatively in cities where there was no rain.[11] These social media studies confirm the hypothesis that mood is contagious, but there is a caveat.

Studies involving millions of posts are analysed with computer algorithms that infer the mood of the users from the presence of certain words that are statistically associated with positive or negative emotions. More positive and fewer negative words in a post is inferred to mean that the user is feeling happy, and the reverse suggests low mood. But how accurate is this? A study that asked hundreds of people to record how they were feeling during the day and then matched these reports with the emotion words they used in Facebook posts, found that neither could be used to predict the other.[12] This suggests that the effects that are detected in studies that sample millions of posts are actually so weak that they disappear into the noise in smaller networks. Individual Facebook users have small social networks.

The same study also used human judges to rate the mood of Facebook posts and these ratings did reveal a correlation between how people said they felt, and the emotions expressed on Facebook. So, better software, perhaps trained by human judgements, would probably show stronger effects of emotional contagion in social networks than current massive-scale studies reveal.

Now, returning to the reason we plunged into social media research in the first place, what does this tell us about the affiliative social function of humour? Taken together with the other evidence we discussed, it supports the hypothesis that humour fosters affiliation by contagiously spreading emotions and that this can happen online as well as in person.

The two social functions of affiliation and superiority can be found working together in the jokes that doctors tell about each other. A French study of medical humour uncovered a whole mother lode of Hobbesian humour that every specialism in the medical profession uses to ridicule the others.[13]

What's a colonoscope? An instrument used by a gastroenterologist that has an arsehole at both ends.

What's the difference between a surgeon and God? God doesn't think he's a surgeon.

What is the difference between a train and a psychiatrist? A train stops when it goes off the rails.

How do you tell a surgeon from an anaesthetist in the operating theatre? The surgeon's gown is stained with blood, the anaesthetist's gown is stained with coffee.

What do you call two orthopaedic surgeons looking at an ECG? A double-blind study.

Though superiority is not the foundation of all humour as many once believed, it is one of its major cultural expressions, with a distinct social function. Humour did not evolve for the purpose of social control, but jokes that ridicule others certainly provide one of its uses. In the United States, racist humour was a bulwark of white supremacy from the era of slavery until the American civil rights movement. White performers who dressed up in blackface and acted the dumb negro played upon and reinforced the prejudices of their white audiences, contributing to the dehumanisation of African-Americans and the defence of slavery. One such performer, Thomas Rice, who invented the character Jim Crow, actually began a performance in 1837 with a speech in which he said: 'I have studied the negro people upon the southern plantations... I effectually proved that negroes are essentially an inferior species of the human family, and they ought to remain slaves.'[14]

The notorious caricatures of Jews that appeared

in Nazi Germany certainly served the same dehumanising function, with genocidal consequences. In recent times, psychologists have extensively investigated the effect on social attitudes of exposing volunteers to racist and sexist humour.[15] These experiments show that such humour does not create prejudice in participants who were previously free of this, but it does reinforce existing sexist and racist attitudes and make people more willing to express prejudices that they might previously have been reluctant to show. Most worryingly, male subjects of such research who liked hostile sexist jokes were more approving of sexual violence towards women after being exposed to sexist humour.[16]

Can humour be used to turn the tables and to disparage and reduce sexist and racist attitudes? There are comedians such as Chris Rock who attempt this, but studies suggest that they are preaching to the choir and that mocking bigotry does not defeat it, however deserving of ridicule bigots might be.[17] Pre-existing prejudice shapes how people perceive jokes that are intended to ridicule their attitudes, so that a joke such as this one, encountered earlier, flops with racists:

What do you call a black aircraft pilot? A pilot.

The research suggests that the humour of

superiority reveals and perhaps magnifies existing social attitudes rather than challenging them.

Laughter can signal ridicule, but of two kinds depending on social context. When the source of the ridicule is dominant, the laughter is superior. When the target of the ridicule is dominant, the laughter is subversive. As often as humour has been used to demonise the oppressed and to justify the superiority of their oppressors, it has also been used subversively, to turn the tables on them and to mock the ruling elite. After the downfall of the Tunisian dictator Ben Ali, this Arab joke circulated on Facebook:

> *Ben Ali goes into a shoe shop to buy a new pair of boots. Scarcely is he through the door than the shop assistant presents him with a pair of boots of exactly the style that he likes in precisely the right size.*
>
> *'How did you know my shoe size?' asks Ali.*
>
> *'Easy,' replies the shop assistant. 'You have been stomping on us for so long, we all know exactly what size boots you wear.'*[18]

George Orwell[19] wrote that 'Every joke is a tiny revolution. Whatever destroys dignity, and brings down the mighty from their seats, preferably with a bump, is funny', which is why despots don't like to be laughed at. During Hitler's regime in Germany, the Nazis had special Peoples' Courts that tried and sentenced to death people who told subversive jokes, like this one, which cost a Marianne K. her life:

Hitler and Goering are on the radio tower in Berlin, looking at the crowds below. Hitler wants to do something to put a smile on the Berliners' faces. So Goering says: 'Why don't you jump?'[20]

According to an underground Soviet-era joke:

A contest for the best political joke was announced.
3rd Prize — 10 years and confiscation of all belongings
2nd Prize — 15 years' solitary confinement
1st Prize — 25 years' imprisonment

The dissident Soviet writer and Nobel-Prize winner Alexander Solzhenitsyn would not have found this joke funny. He served eight years in a Siberian prison camp for making a sarcastic remark about Stalin in a private letter. A joke of the time went:

What is the difference between Teddy Roosevelt and Joseph Stalin? Roosevelt collected the jokes that people made about him. Stalin collected the people who made jokes about him.[21]

Stalin sent millions of people to the labour camps of the Gulag for telling political jokes.

Who built the White Sea Canal?

The right bank was built by those who told jokes.

And the left bank?

By those who listened.

Stalin enjoyed anti-Semitic jokes and made jokes about his own brutal reputation. Only one man in his inner circle, Karl Radek, who happened to be Jewish, is known to have joked to Stalin's face and survived, at least for a while. One of Radek's innumerable jokes was the now well-known definition of the difference between capitalism and communism:

Capitalism is the exploitation of man by man, and communism is the reverse.

But even Radek eventually disappeared.

What were Radek's last words before he committed suicide?

Don't shoot!

Totalitarianism was so enduring that it generated

enough subversive humour for the entire 72-year history of the Soviet Union to be retold through jokes. In a book entitled *Hammer and Tickle,* Ben Lewis reveals that the jokes told during Soviet times were often re-versioned from anti-Czarist jokes of the pre-revolutionary era.[22] What is more, the same jokes were told about the communist regimes in different countries of the Eastern Bloc, though in each country they were regarded as home inventions that illustrated the native wit of the Czechs, the Bulgarians, the Hungarians or whoever was telling them.

An enduring theme from Czarist times and right across the Soviet Union and its satellites is the obviously Jewish identity of the comic protagonist:

'Comrade Rabinowitz, why didn't you attend the last meeting of the Communist party?'

'No one told me that it would be the last one! If I had known that, I would have brought my whole family.'[23]

Even after Stalin died, Jews remained frequent protagonists in political jokes:

The Soviet Union decides that it should commemorate Lenin's stay in Zurich, where he spent a year before returning to Russia to lead the Bolshevik revolution in 1917. They decide to commission an artist from Leningrad to paint a picture to be called 'Lenin in Zurich'. Some older members of the party object that the painter is Jewish, but it is decided that in the modern Soviet Union this is no longer a problem. Twelve months go by, during which time the artist works in secrecy on his masterwork. The day of the unveiling arrives. The painting is hung, draped with a cloth, in the Kremlin. The Red Army Band plays, the Red Army Choir sings, a commissar gives a long speech, and then the sash is pulled. As the cloth falls away there is a gasp of horror from the crowd. In outrage and disbelief they stare at the painting, which depicts Lenin's wife, naked, in bed with Trotsky, who is wearing nothing but his pince-nez.

'Where is Comrade Lenin?' the commissar asks, scarcely able to control himself.

The painter steps forwards and says, 'In this painting, Lenin is in Zurich.'

Humour has such a particular character and importance in Jewish life that Jewish jokes are a natural medium through which to explore the cultural significance of humour. What is a Jewish joke? Why are there so many Jewish comedians? Where does this font of wit come from?

Many writers have tried to define what is so

characteristically Jewish about Jewish humour, but without any obvious success. Sigmund Freud collected Jewish jokes and used them to illustrate his treatise on *Jokes and their Relation to the Unconscious*.[24] Most of Freud's jokes are not particularly Jewish. For example:

> *Two Jews met in the neighbourhood of the bath house. 'Have you taken a bath?' asked one of them.*
> *'What?' asked the other in return, 'is there one missing?'*

And:

> *Two Jews were discussing baths. 'I have a bath every year,' said one of them, 'whether I need one or not.'*

Freud described this joke as rather coarse, betraying a prudishness that probably explains why, although he was academically interested in sex, he totally ignored the very rich vein of Jewish sex jokes. It should be no surprise to find that defining the Jewish joke is difficult. Why wouldn't it be? On most topics it's fair to say, 'Two Jews, three opinions.'

> *A new rabbi is appointed to the synagogue and finds that every Shabbat (Sabbath) the congregation is divided into two camps who spend half the*

service shouting at each other. In one camp are the congregants, who recite the holy Shema while standing up, and in the other are those who believe that it must be recited while seated. The rabbi gets fed up with the disruption and invites a representative of the standers and a spokesperson for the sitters to meet him. He also asks along the oldest member of the congregation to advise on tradition and he asks him: 'Is it traditional in this Shul (synagogue) to recite the Shema while standing up?'

'No, that's not the tradition,' replies the old man.

'Aha!' says the sitter. 'You see, we are right!'

'Well,' says the rabbi to the old man, 'Is it traditional in this Shul to recite the Shema while sitting down?'

'No, that's not the tradition,' replies the old man.

'How can that be?' says the rabbi. 'It's not the tradition to stand and it's not the tradition to sit. Everyone is just going to carry on arguing!'

'<u>That's</u> the tradition,' says the old man.

But, argumentativeness is not confined to Jews by any means. Protestants seem to be quite susceptible to it, too. Near where I live in Edinburgh, churches belonging to four different Protestant denominations face off against each other at a crossroads that locals call Holy Corner. In his book *The Mirth of Nations*, the leading scholar of jokes and culture Christie Davies finds many parallels between Jewish humour and that of the Scots in the late 19th and early 20th centuries when the four churches at Holy Corner were built.[25] The similarities, he argues, are due to Scots being a minority in an English-dominated world. So, this Jewish joke could just as easily be Presbyterian:

> *Goldberg is shipwrecked and washes up on a desert island, where he is the only survivor. He spends many years alone on the island until one day a passing cruise ship spots his smoke signals and sends a boat ashore to rescue him. His rescuer is curious and asks, 'What is that hut over there?'*
>
> *'Oh, that is my shul,' Goldberg replies proudly. 'I built it myself.'*
>
> *'And what about that hut over there, then?'*
>
> *'That!? That is the shul that I wouldn't be seen dead in!'*

This same joke can also be found in a Welsh version where the shuls are non-conformist chapels. It doesn't matter whether the joke was originally Jewish or not. The tendency it satirises is clearly not exclusively Jewish. Even so, is any other religion as obsessed with the interpretation of biblical law, or as keen to make it more challenging to observe? Take the Jewish dietary laws, for example:

God: Thou shalt not stew the kid in its mother's milk. It's cruel.
Rabbi: Milk must never be eaten with meat, then.
God: Thou shalt not stew the kid in its mother's milk.
Rabbi: We must use separate plates for milk and for meat.
God: Thou shall not stew the kid in its mother's milk.
Rabbi: If the wrong plate is used, it must be buried in the ground for six weeks.
God: Oh, please yourself!

Philosopher Ted Cohen believes that the tradition of biblical study and exegesis created the culturally fertile ground in which the Jewish sense of humour developed.[26] *The Soncino Chumash*, which was the daily study of my father in his later years, is an edition of the *Five Books of Moses* embalmed in footnotes, annotations, arguments and editorial embellishments much longer than its biblical core. Yet the *Soncino* is a mere pocket edition by comparison with the Babylonian Talmud which comes in at 6,200 pages.

How does a religious obsession with biblical

interpretation become a seedbed for humour? Leonard Cohen gave us a clue in a quip made on the come-back tour that he gave after his manager stole all his retirement savings:

> *'I was 60 years old [when last on tour] — just a kid with a crazy dream. Since then I've taken a lot of Prozac, Paxil, Wellbutrin, Effexor, Ritalin, Focalin … I've also studied deeply in the philosophies and religions, but cheerfulness kept breaking through.'*

When life is deadly serious and you are staring penury and finitude in the face, do you laugh, or do you cry? Both.

If Ted Cohen is right that the Old Testament is the seedbed of Jewish humour, it is strangely light on laughs. Abraham and Sarah named their son Isaac, which means laughter in Hebrew, but that's about it. The joke appears to have been that Isaac was born when his father was 100 and his mother 90. Not the funniest punch line, especially for poor Sarah.

If Jewish humour is not to be found in the Bible then, where does it come from? A stronger case can be made that it is the humour of the outsider. As Tom Lehrer observes in his song 'National Brotherhood Week':

Oh, the Protestants hate the Catholics,
And the Catholics hate the Protestants,
And the Hindus hate the Muslims,
And everybody hates the Jews.

Jewish humour is often self-deprecatory, which is perhaps another clue to its outsider status. Groucho Marx expressed this in the ultimate outsider joke: 'I don't care to belong to any club that would have me as a member.' Jews, even of Groucho's celebrity, were excluded from country clubs. Groucho's young son was rejected for membership of a country club where he wanted to swim. Groucho wrote to the club asking whether, since his son was only half Jewish, he would be allowed to go into the club's pool just up to his waist?[27]

In contrast, when presenting himself at a customs post, Oscar Wilde said 'I have nothing to declare but my genius.' Alexandre Dumas reported that at a dinner party for politicians 'I would have been bored if I had not been present myself.' Wilde was gay and Dumas was black, and both were highly celebrated. There are outsiders and there are Jewish outsiders.

'Did you hear that Jews sank the Titanic?'

'The Jews? I thought it was an iceberg.'

'Iceberg, Goldberg, Rosenberg — they're all the same.'

The difference between Jews and other outsiders is the conspiracy theories.

> *Rabbi Altmann and his secretary were sitting in a Berlin coffee house in 1935.*
>
> *'Herr Altmann,' said his secretary, 'I notice that you are reading* Der Stürmer*! A Nazi libel sheet. Are you some kind of masochist?'*
>
> *'On the contrary, Frau Epstein. When I used to read the Jewish newspapers, they were full of depressing news about pogroms, riots in Palestine, and people leaving the faith in America. But now I read* Der Stürmer, *I see that we Jews control all the banks, dominate the arts and sciences, and are on the verge of taking over the entire world, and I feel so much better.'*

Here, surely, is where Jewishness finds such affinity with humour. To be Jewish is to have to accommodate two polar opposites at once — to be both chosen and to be picked-upon. Living with and resolving this incongruity produces humour.

The test of this hypothesis is Israel. On the one hand, you might imagine that the Jewish state would be a rich source of typically Jewish jokes. However, since Israel is the one place where Jews are not outsiders, it may ironically be the last place to look for humour that's characteristically Jewish. Here is the ultimate proof of Israeli self-confidence:

It is 1985. Israel's economy is in bad shape and the Knesset holds a special session to discuss what to do. After hours of debate, one member stands up and says, 'Quiet, everyone, I've got it, the solution to all our problems.'

'What?' inquires the Knesset's Speaker.

'We'll declare war on the United States.'

'You're nuts! That's crazy! What, are you mad?'

'Wait, hear me out. We declare war and we lose. The United States then does what she always does when she defeats a country and invests, like with the Marshall Plan. We'll get new roads, airports, docks, schools, hospitals, factories, and even food aid. Our problems will be over.'

'Sure,' says another member, '<u>if</u> we lose.'

For the most part, Israeli jokes are the same as those found elsewhere and are not obviously Jewish. Versions of the joke about the stupidity of the Israeli politician David Levi (p. 26–7) are, after all, found in many countries; in Israel, this joke is political, not Jewish. Rabbi Joseph Telushkin, an American authority and great interpreter of Jewish humour has written:

> There is not a great deal of humour being created in Israel, and most of what exists is not very funny at least to non-Israelis. Because people in power are able to deal with their problems directly, they have no need to settle for the personal gratification of a sharp put-down or witticism.[28]

The example of Israel supports the hypothesis that Jewish humour as heard in the diaspora is outsiders' humour. Devorah Baum, another writer on Jewish humour,[29] observes that, in Israel, the role of the outsider has passed to the Palestinians.

Jewish humour demonstrates something important about joke culture that may apply more generally. Jewish humour contains two distinct elements — a culture-specific element that is affiliative in social function, and a second that is subversive. The affiliative element of Jewish humour reflects the cultural and religious particulars of Judaism, like the joke about the dietary laws. Such particulars are present in the jokes of all cultures. For example, among the deaf community there are a great many jokes about cochlear implants of which hearing people have no experience, making them unable to share the joke. Deaf jokes told with sign language often contain punning hand gestures — the visual equivalent of homophones.[30] The equivalent in

Jewish humour is the jokes that pun in Yiddish.

The second element in Jewish humour, with a subversive social function, reflects the outsider status of Jews in the diaspora. The deaf community also has outsider status and even shares some of the same jokes, like this one:

> *A Catholic priest goes to the barber's. When he is finished, the barber says, 'As a good Catholic myself, of course I'm not going to charge you for the haircut, Father.' The priest goes away and returns with the gift of a rosary for the barber.*
>
> *An Anglican priest goes to the same barber, who again cuts the religious man's hair for free. The Anglican returns with a box of chocolates that he gives to the barber.*
>
> *Finally, a Rabbi goes for a haircut. The barber says, 'Rabbi, I respect all men of God, so I am not going to charge you for this haircut.' The Rabbi goes away and returns with another Rabbi.*

In the deaf version of this joke, the first two visitors to the barber's are a man in a wheelchair and a blind man. Both return to give the barber a present, but after the deaf man receives his free haircut, he returns with all his deaf friends. This joke is affiliative as well as subversive for deaf people because they recognise the habit of sharing good things in their community, and understand how rare it is that they can get the better of hearing people.[31]

To what extent do the different social functions of laughter vary between joke cultures? The reward function of laughter is common to all humour and research shows that this creates increased affiliation. These are therefore the foundation of all joke cultures rather than differentiating among them. This leaves dominance and subversive social functions to do the heavy lifting in defining a joke culture. Applying this to Jewish humour: diaspora and Israeli humour are both Jewish in origin, but the first belongs to outsiders and is subversive, while the latter is not, which is why they are different.

Another conclusion to be drawn is that since Israeli society is a young scion of an ancient people, the difference between Israeli and Jewish humour demonstrates how rapidly a joke culture can change with circumstances.

Jewish humour is abundant and familiar, but how does it compare with a different culture such as Chinese, which is also ancient, yet isolated until relatively recently? There are of course Jewish jokes about this:

Sam Cohen greets his buddy and near namesake

Sam Chen at their favourite Chinese restaurant to celebrate Chinese New Year. 'Our calendar is more than 4,600 years old, you know,' says Chen.

'No big deal,' says Cohen, 'Our Hebrew calendar is more than 5,700 years old.'

'Really? What did you eat for the first thousand years?'

A Chinese couple have just dined in a Jewish restaurant in New York. 'Jewish food is all very well,' says one to the other, 'but after three days you feel hungry again.'

So much for the differences, but there are similarities, too. Chinese culture particularly favours self-deprecating jokes and affiliative humour. Yu-Chen Chan and her colleagues In Taiwan used fMRI scans to see how volunteer's brains reacted to sets of jokes that differed only in whether the butt of the joke was the joke teller (self-deprecating) or another person (aggressive) or whether it flattered the listener (affiliative) or the joke teller (self-aggrandising).[32] For example, one set of jokes was:

'If each of my admirers were a strand of hair, I would be bald.' (Self-deprecating)

'If each of your admirers were a strand of hair, you would be bald.' (Aggressive)

'If each of your admirers were a strand of hair, you would need two heads.' (Affiliative)

'If each of my admirers were a strand of hair, I would need two heads.' (Self-aggrandising)

You can of course just ask people which jokes they find funniest and volunteers in this study did say that these one-liners were funny, but the fMRI scans give a more objective response by which the different jokes' conditions can be compared. The scans revealed that the Chinese subjects of the study appreciated self-deprecating and affiliative humour more than humour that was aggressive or self-aggrandising. It mattered less whether the butt of the joke was the teller or the listener than whether the joke was affiliative / self-deprecating or not.

The way that people react to aggressive humour varies between different cultures. For example, a study comparing attitudes to humour in Taiwan with that of German-speakers in Switzerland found that people in Taiwan were more afraid of being laughed at than the Swiss.[33] Just as well since, as we all know, the Swiss in their nuclear bunkers will have the last laugh:

Who will be left after a global nuclear war? Just cockroaches and the Swiss.

Christie Davies has suggested that in more hierarchical societies such as China and Japan, the fear of losing face is greater than in more individualistic, Western ones and that this is reflected in the fear of being the butt of a joke. Maybe so, but authority does not escape ridicule anywhere and probably never has.

In 18th-century Japan a comic style of haiku poem called *Senryū* was invented by a poet writing under the pen name Karai Senryū (1718–1790). He began a public competition for the best humorous observations made in the haiku form, which consists of 17 syllables falling in three lines in the pattern five-seven-five. In 1767, an incredible 140,000 *Senryū* were submitted to the competition held that year. The winning *Senryū* were published in anthologies and included such examples as this comment on graft:[34]

The civil servant's baby
Is very good at learning
How to grasp.

And this affiliative *Senryū* which refers to the social obligation for a man to join in with his companions' laughter, even if he is unable to drink alcohol with them:

The bloke not drinking,
Laughs now and then.
That's all.

In 1966, when the Chinese leader Mao Zedong was 72, a wall newspaper appeared with a photo captioned 'Our great leader Chairman Mao has a good swim in the Yangtze River.' The following joke written alongside earned its author six years in jail:[35]

Chairman Mao swims in the Yangtze at such an advanced age. He has installed a propeller underneath.

Today, even in the heavily censored arena of Chinese cyberspace, people manage to share video parodies and political memes. In 2013, a meme that mocked President Xi Jinping for his portly resemblance to Winnie the Pooh began spreading online. A year later, a photo of Xi shaking hands with the Japanese Prime Minister Shinzō Abe began circulating alongside an uncannily similar image from a Disney cartoon of Pooh holding paws with the donkey Eeyore, who bore the same lugubrious expression as the Japanese Prime Minister. In 2015, a similar meme became the most censored image of the year and in 2018 the Chinese Government banned the latest Winnie the Pooh movie.[36]

What do Winnie the Pooh and Atilla the Hun have in common? The same middle name.

But I digress.

Whatever the cultural differences between nations, it is difficult to escape the conclusion that people everywhere use jokes — even sometimes the same ones — to deflate authority and kick back against oppression in whatever medium provides them with the opportunity, be that *Senryū*, memes, cartoons, or graffiti. When the four social functions of laughter are considered, the differences among cultures seem smaller and less enduring than the fundamental similarities.

More than 150 jokes ago, I asked 'What good is laughter?' That question can be answered in different ways and we've seen that when a sociologist asks it, she wants to know what laughter does in society. Laughter is a biological feature and so are the emotions of reward and social affiliation that arise from it, so I approached the question as an evolutionary one. This starts with the fundamental question, 'Why does laughter exist in the first place?' You might call this a Darwinian approach to laughter, especially since Darwin himself considered the question. Darwinian explanations seek not only to understand how a biological feature could have evolved through natural selection, but also how it got going. We have seen how laughter probably evolved as a play vocalisation tens of millions of years ago. Before that, the original laugh that became a play vocalisation may have been an all-clear signal.

At some point in our more recent evolutionary history, humour became a trigger for the play vocalisation and we began to laugh at benign incongruities. The last sentence should be written in neon lights 10 metres tall because it represents the culmination of more than 2,000 years of thinking by some of the brightest minds on the planet. Aristotle, were he alive today, might well lead a chorus of philosophers in a shout of 'I told you so,' but the difference now is that we have the evidence of how it could have happened.

The puzzle, you will remember, is not that we laugh at errors and incongruities, but that we only laugh at trivial ones. We have this massive brain that got us through a disaster in Africa that 180,000 years ago brought our species near to extinction and is now propelling us towards a global population of 10 billion, and what do we do with the supercomputer in our frontal cortex? Crack jokes. It turns out that this is not the paradox it seems, because jokes are advertisements for intellectual ability. Let me repeat the refrain of Chapter 6: laughter is not about survival but seduction, not about defence but display. Sexual selection is the key: humour is a demonstration that you have the wits to woo.

Lily Tomlin deserves the last word because she nailed it when she said:

> *Instead of working for the survival of the fittest, we should be working for the survival of the wittiest — then we can all die laughing.*[37]

Acknowledgements

I am, as ever, grateful to Rissa de la Paz for her uncompromising, critical comments on the manuscript at every stage of its development. Prof. Adrian Moore, equally expert in humour and philosophy, and Prof. Daniel Nettle, expert in evolutionary psychology, spent unrewarded hours reading the whole book, and I am grateful to them both. I thank my friends and fellow writers in Edinburgh's Stranger Than Fiction group for their comments and encouragement when the book was just a sketch: Vin Arthey, Rachel Blanche, Maria Chamberlain, George Davidson, Murray Earle, Alex Owen-Hill, Guthrie Stewart, and Anne Wellman. Eleanor Birne and Philip Gwyn Jones provided useful comments on the final draft, and Molly Slight guided it through to publication.

Notes

Chapter One: Comedy and Error

1 Bate, J. & Rasmussen, E. (2007), eds. *The RSC Shakespeare: the complete works.* Basingstoke: Palgrave Macmillan.

2 Provine, R.R. (2001). *Laughter: a scientific investigation.* London: Penguin Books.

3 Aristotle, *The Poetics.* Translated by S.H.Butcher. Project Gutenberg. https://www.gutenberg.org/files/1974/1974-h/1974-h.htm [Accessed 28 December 2018].

4 Ghose, I. (2008). *Shakespeare and Laughter: a cultural history.* Manchester: Manchester University Press.

5 Raskin, V. (2008). Theory of Humor and Practice of Humor Research: editor's notes and thoughts. In V. Raskin, ed. *The Primer of Humor Research.* Berlin and Boston, MA: De Gruyter, Inc.

6 McGhee, P.E. and Goldstein, J.H. (1983). *Handbook of Humor Research: volume 1: basic issues.* Berlin: Springer-Verlag.

7 Dupont, S., *et al.* (2016). Laughter Research: A Review of the ILHAIRE Project. In: A. Esposito and L. C. Jain, eds. *Toward Robotic Socially Believable Behaving Systems, Vol I: Modeling Emotions.* Intelligent Systems Reference Library 105. New York, NY: Springer Publishing. pp. 147–81.

8 Ken Dodd, *Night Waves*, BBC Radio 3. First broadcast June 2012.
9 Provine. *Laughter*, op. cit.
10 Ekman. P. (1999) ed. *Charles Darwin: The Expression of the Emotions in Man and Animals.* London: Fontana Press.
11 Barry, J.M., Blank, B. & Boileau, M. (1980). Nocturnal penile tumescence monitoring with stamps. *Urology*, 15, 171–172.
12 Weems, S. (2014). *Ha!: The Science of When We Laugh and Why*. New York, NY: Basic Books.

Chapter Two: Humour and Mind

1 Sherrin, N. (2005). *Oxford Dictionary of Humorous Quotations*. Oxford: Oxford University Press.
2 Carr, J. & Greeves, L. (2007). *The Naked Jape: uncovering the hidden world of jokes*. London: Penguin Books.
3 Saxe, J.G. (1876) *Poems*. Boston, MA: J.R. Osgood and Co.
4 https://tickets.edfringe.com/whats-on#fq=venue_name%3A%22Pleasance%20Courtyard%22&fq=-category%3A(%22Comedy%22)&fq=subcategories%3A(%22Satire%22)&q=*%3A* [Accessed 28 December 2018].
5 Aristotle, *The Poetics*. Translated by S.H. Butcher. Project Gutenberg. https://www.gutenberg.org/files/1974/1974-h/1974-h.htm [Accessed 28 December 2018].
6 Crompton, D. (2013). *A Funny Thing Happened on the Way to the Forum: the world's oldest joke book.* London: Michael O'Mara.
7 Hobbes, T. (1840). *Human Nature.* London: Bohn.
8 Hurley, M.M., Dennett, D.C. & Adams, R.B. (2011). *Inside Jokes: using humor to reverse-engineer the mind.* Cambridge, MA: MIT Press.

9 Rebecca West, quoted in Sherrin. *Humorous Quotations*. op. cit.
10 Jarski, R. (2004). *The Funniest Thing You Never Said: the ultimate collection of humorous quotations*. London: Ebury Press.
11 Quoted in: Hurley, Dennett & Adams. *Inside Jokes*. op. cit.
12 Greengross, G., Martin, R.A. & Miller, G. (2012). Personality Traits, Intelligence, Humor Styles, and Humor Production Ability of Professional Stand-up Comedians Compared to College Students. *Psychology of Aesthetics Creativity and the Arts*, 6, 74–82.
13 Arnott, S. & Haskins, M. (2004). *Man Walks into a Bar: the ultimate collection of jokes and one-liners*. London: Ebury Press.
14 Told by Richard Wiseman, University of Hertfordshire, to *The Observer*, 29 December 2013 (p.23).
15 Hurley, Dennett & Adams. *Inside Jokes*. op. cit.
16 'One morning I shot an elephant in my pyjamas' Groucho Marx, *Animal Crackers* (1930).
17 https://www.chortle.co.uk/news/2014/04/03/19917/tim_vine_retakes_most_jokes_in_an_hour_record [Accessed 14 July 2019].
18 Vine, T. (2010). *The Biggest Ever Tim Vine Joke Book*. London: Cornerstone.
19 Ekman. P. (1999) ed. *Charles Darwin: The Expression of the Emotions in Man and Animals*. London: Fontana Press.
20 Attardo, S. (2008). A primer for the linguistics of humour. In: V. Raskin, ed. *The Primer of Humor Research*. Berlin & Boston, MA: De Gruyter, Inc.; Rapp, A. (1949). A phylogenetic theory of wit and humor. *Journal of Social Psychology*, 30, A81–A96.
21 Kant, I. (1790). *Critique of Judgment*. Trans. W.S. Pluhar. (1987) Indianapolis, Indiana: Hackett Publishing Company.

22 Gumbel, A. (2004). Obituary: Professor Sidney Morgenbesser. *Independent*, London. 6 August 2004.
23 Adamson, J. (1974). *Groucho, Harpo, Chico, and sometimes Zeppo: a history of the Marx Brothers and a satire on the rest of the world.* London: Coronet Books.
24 BBCNews (2018). Airline spells own name wrong on plane. https://www.bbc.com/news/world-asia-45572275. [Accessed 14 January 2019].
25 Chan, Y.C. *et al.* (2013). Towards a Neural Circuit Model of Verbal Humor Processing: an fMRI study of the neural substrates of incongruity detection and resolution. *Neuroimage*, 66, 169–176.
26 Martin, R.A. & Ford, T.E. (2018). The Physiological Psychology of Humor. In: Martin, R.A. & Ford, T.E., eds. *The Psychology of Humor: an integrative approach.* (Second Edition) London: Academic Press; Nakamura, T. *et al.* (2018). The role of the amygdala in incongruity resolution: the case of humor comprehension. *Social Neuroscience*, 13, 553–565.

Chapter Three: Song and Dance

1 https://youtu.be/nDZZEfrRbdw [Accessed 30 April 2020].
2 https://en.wikipedia.org/wiki/Goldberg_Variations#-Variatio_30._a_1_Clav._Quodlibet [Accessed 30 April 2020].
3 Eriksen, A.O. (2016). A Taxonomy of Humor in Instrumental Music. *Journal of Musicological Research*, 35, 233–263. 10.1080/01411896.2016.1193418.
4 Huron, D. (2004). Music-engendered Laughter: An analysis of humor devices in PDQ Bach. In: S.D. Lipscombe, R. Ashley, R.O. Gjerdingen & P. Webster, eds. *Proceedings of the 8th International Conference on*

Music Perception and Cognition, Evanston, Illinois. pp. 700–704.

5 Hashimoto, T., Hirata, Y. & Kuriki, S. (2000). Auditory Cortex Responds in 100 ms to Incongruity of Melody. *Neuroreport*, 11, 2799–2801.

6 Halpern, A.R. *et al.* (2017). That Note Sounds Wrong!: age-related effects in processing of musical expectation. *Brain and Cognition*, 113, 1–9.

7 Sutton, R.A. (1997) Humor, Mischief, and Aesthetics in Javanese Gamelan Music. *Journal of Musicology*, 15, 390–415.

8 Nerhardt, G. (1970). Humor and inclination to laugh — emotional reactions to stimuli of different divergence from a range of expectancy. *Scandinavian Journal of Psychology*, 11, 185–195; Deckers, L. & Kizer, P. (1974). Note on weight discrepancy and humor. *Journal of Psychology*, 86, 309–312.

9 Lieberman, P. (2015). Language Did Not Spring Forth 100,000 Years Ago. *Plos Biology*, 13. 10.1371/journal. pbio.1002064.; Dediu, D. & Levinson, S. C. (2018). Neanderthal language revisited: not only us. *Current Opinion in Behavioral Sciences*, 21, 49–55.

10 Ken Dodd delivering his three-legged chicken joke: https://youtu.be/KuMLHytm_O0 [Accessed 30 April 2020].

11 Gimbel, S. (2018). *Isn't That Clever: a philosophical account of humor*. New York and London: Routledge.

12 Hull, R., Tosun, S. & Vaid, J. (2017). What's so funny? Modelling incongruity in humour production. *Cognition & Emotion*, 31, 484–499.

13 Hempelmann, C.F. (2008). Computational humor: Beyond the pun?. In: V. Raskin, ed. *The Primer of Humor Research*. Berlin & Boston, MA: De Gruyter, Inc.

14 Strapparava, C., Stock, O. & Mihalcea, R. (2011)

Computational Humour. Emotion-Oriented Systems: Cognitive Technologies (eds P. Petta, C. Pelachaud & R. Cowie), pp. 609–634. Springer-Verlag, Berlin & Heidelberg.

15 https://inews.co.uk/culture/100-best-jokes-one-liners-edinburgh-fringe/ [Accessed 27 April 2019].

16 Hempelmann, C.F. (2008). Computational Humor: beyond the pun?. In Raskin, *Primer*, op. cit.

17 These are from http://joking.abdn.ac.uk/jokebook.shtml. [Accessed 11 February 2019].

18 Brooke-Taylor, T. *et al.* (2017). *The Complete Uxbridge English Dictionary*. London: Windmill Books.

19 https://en.wikiquote.org/wiki/Sidney_Morgenbesser. [Accessed 27 April 2019]

20 Gumbel, A. (2004). Obituary: Professor Sidney Morgenbesser. *Independent*, London. 6 August 2004.

21 Hurley, M.M., Dennett, D.C. & Adams, R.B. (2011). *Inside jokes: using humor to reverse-engineer the mind.* Cambridge, MA: MIT Press.

22 Sherrin, N. (2005). *Oxford Dictionary of Humorous Quotations*. Oxford: Oxford University Press.

23 McCrae, R.R. & John, O.P. (1992). An introduction to the 5-factor model and its applications. *Journal of Personality*, 60, 175–215.

24 Berger, P. *et al.* (2018). Personality modulates amygdala and insula connectivity during humor appreciation: An event-related fMRI study. *Social Neuroscience*, 13, 756–768.

25 Martin, R.A. & Ford, T.E. (2018). The Personality Psychology of Humor. In: R. A. Martin & T. E. Ford, eds. *The Psychology of Humor*. Second Edition. London: Academic Press. pp. 99–140.

26 Schweizer, B. & Ott, K.H. (2016). Faith and laughter: Do atheists and practicing Christians have different senses of humor? *Humor-International Journal of Humor*

Research, 29, 413–438; Wiseman, R. (2008). *Quirkology: The curious science of everyday lives*. London: Pan Books.

27 Gabora, L. & Kitto, K. (2017). Toward a Quantum Theory of Humor. *Frontiers in Physics*, 4, Article #53.

Chapter Four: Tickle and Play

1 Darwin, C. (1999) *Charles Darwin: The Expression of the Emotions in Man and Animals*. (ed P. Ekman). London: Fontana Press.

2 Davila-Ross, M., Owren, M. J. & Zimmermann, E. (2014). The evolution of laughter in great apes and humans. *Communicative & Integrative Biology*, 3, 191–194.

3 Bard, K.A. *et al.* (2014). Gestures and social-emotional communicative development in chimpanzee infants. *American Journal of Primatology*, 76, 14–29.

4 Bryant, G.A. & Aktipis, C.A. (2014). The animal nature of spontaneous human laughter. *Evolution and Human Behavior*, 35, 327–335.

5 Lavan, N. *et al.* (2018). Impoverished encoding of speaker identity in spontaneous laughter. *Evolution and Human Behavior*, 39, 139–145.

6 Maynard Smith, J. & Harper, D. (2003). *Animal signals*. Oxford: Oxford University Press.

7 Ramachandran, V.S. (1998). The neurology and evolution of humor, laughter, and smiling: the false alarm theory. *Medical Hypotheses*, 51, 351–354.

8 Darwin, C. *The Expression of the Emotions*, op. cit.

9 Quoted in: Harris, C.R. (1999). The mystery of ticklish laughter. *American Scientist*, 87, 344–351.

10 Blakemore, S.-J., Frith, C.D. & Wolpert, D.M. (1999). Spatio-Temporal Prediction Modulates the Perception of Self-Produced Stimuli. *Journal of Cognitive Neuroscience*, 11, 551.

11 Blakemore, S.J. *et al.* (2000). The perception of self-produced sensory stimuli in patients with auditory hallucinations and passivity experiences: evidence for a breakdown in self-monitoring. *Psychological Medicine*, 30, 1131–1139.

12 Gary Shandling, quoted in Jarski, R. (2004). *The Funniest Thing You Never Said: The ultimate collection of humorous quotations*. London: Ebury Press.

13 Provine, R.R. (2001). *Laughter: a scientific investigation*. London: Penguin.

14 Wöhr, M. (2018). Ultrasonic communication in rats: appetitive 50-kHz ultrasonic vocalizations as social contact calls. *Behavioral Ecology and Sociobiology*, 72. DOI: 10.1007/s00265–017–2427–9.

15 Martin, R.A. & Ford, T.E. (2018). The Physiological Psychology of Humor. In: R.A. Martin, & T.E. Ford, eds. *The Psychology of Humor: an integrative approach*. Second Edition. London: Academic Press. pp. 174–204. Watch a video of rats laughing here: https://youtu.be/j-admRGFVNM.

16 Reinhold, A.S., J. I. Sanguinetti-Scheck, K. Hartmann and M. Brecht (2019). Behavioural and neural correlates of hide-and-seek in rats. *Science*, 365, (6458): 1180–1183.

17 Carr, J. & Greeves, L. (2007). *The Naked Jape: uncovering the hidden world of jokes.* London: Penguin.

18 Caeiro, C., Guo, K. & Mills, D. (2017). Dogs and humans respond to emotionally competent stimuli by producing different facial actions. *Scientific Reports,* 7. 10.1038/s41598–017–15091–4.

19 Wang, K. (2018). Quantitative and functional posttranslational modification proteomics reveals that TREPH1 plays a role in plant touch-delayed bolting. *Proceedings of the National Academy of Sciences of the United States of America*, 115, E10265-E10274.

20 Weisfeld, G.E. (1993). The adaptive value of humor and laughter. *Ethology and Sociobiology*, 14, 141–169.

21 Silvertown, J.W. (2017). *Dinner with Darwin: food, drink, and evolution*. Chicago, IL: University of Chicago Press.

22 Gold, K.C. & Watson, L.M. (2018). In memoriam: Koko, a remarkable gorilla. *American Journal of Primatology*, 80. e22930.

23 Koko the Gorilla meets Robin Williams https://youtu.be/vOVS9zotSqM. [Accessed 9 March 2019].

24 McGhee, P. (2018). Chimpanzee and gorilla humor: progressive emergence from origins in the wild to captivity to sign language learning. *Humour*, 31, 405–449. All the Koko information in this section come from this source unless cited otherwise.

25 Roberts, M. (2018). How Koko the gorilla spoke to us. *Washington Post*. 21 June 2018.

26 Mirsky, S. (1998). Gorilla in our midst [Excerpts from & ironic interpretation of online conversation with Koko the gorilla]. *Scientific American*, 279, 28.

27 Hobaiter, C. & Byrne, R.W. (2014). The meanings of chimpanzee gestures. *Current Biology*, 24, 1596–1600.

28 Kühl, H.S. *et al.* (2019). Human impact erodes chimpanzee behavioral diversity. *Science*, 363, 1453–1455.

29 Time Tree of Life: http://www.timetree.org/ [Accessed 9 March 2019]; Besenbacher, S. *et. al.* (2019). Direct estimation of mutations in great apes reconciles phylogenetic dating. *Nature Ecology & Evolution*, 3, 286–292. 10.1038/s41559–018–0778-x.

30 Dunbar, R.I.M. (2012). Bridging the bonding gap: the transition from primates to humans. *Philosophical Transactions of the Royal Society B-Biological Sciences*, 367, 1837–1846. 10.1098/rstb.2011.0217.

31 Manninen, S., *et al.* (2017). Social Laughter Triggers

Endogenous Opioid Release in Humans. *Journal of Neuroscience*, 37, 6125–6131.

Chapter Five: Smile and Wave

1 LaFrance, M. (2013). *Why smile?: The science behind facial expressions*. New York, NY: W.W. Norton.
2 Rychlowska, M., Jack, R.E., Garrod, O.G.B., Schyns, P.G., Martin, J.D. & Niedenthal, P.M. (2017). Functional Smiles: tools for love, sympathy, and war. *Psychological Science*, 28, 1259–1270. 10.1177_0956797617706082.
3 Ruiz-Belda, M.A., Fernandez-Dols, J.M., Carrera, P. & Barchard, K. (2003). Spontaneous facial expressions of happy bowlers and soccer fans. *Cognition & Emotion*, 17, 315–326; Crivelli, C. & Fridlund, A.J. (2018). Facial Displays Are Tools for Social Influence. *Trends in Cognitive Sciences*, 22, 388–399.
4 Owren, M.J. & Bachorowski, J.A. (2001). The evolution of emotional expression: a 'selfish-gene' account of smiling and laughter in early hominids and humans. In: T. J. Mayne & G. A. Bonanno, eds. *Emotions: Current issues and future directions*. New York: Guilford Press. pp. 152–191.; Ramachandran, V.S. (1998). The neurology and evolution of humor, laughter, and smiling: the false alarm theory. *Medical Hypotheses*, 51, 351–354.
5 Martin, J., Rychlowska, M., Wood, A. & Niedenthal, P. (2017), Smiles as Multipurpose Social Signals. *Trends in Cognitive Sciences*, 21, 864–877.

Chapter Six: Laughter and Sex

1 Hurley, M.M., Dennett, D. C. & Adams, R.B. (2011). *Inside Jokes: using humor to reverse-engineer the mind*. Cambridge, MA: MIT Press.
2 Greengross, G. & Mankoff, R. (2012). Book Review: Inside

'Inside Jokes': the hidden side of humor. *Evolutionary Psychology*, 10, DOI: 10.1177/147470491201000305.

3 Miller, G. (2001). *The mating mind: how sexual choice shaped the evolution of human nature*. London: Vintage Books.

4 Darwin, C. (1901). *The Descent of Man, and Selection in Relation to Sex*. London: J. Murray.

5 Puts, D. (2016). Human sexual selection. *Current Opinion in Psychology*, 7, 28–32.

6 Dorothy Parker quotes are from Sherrin, N. (2005). *Oxford dictionary of humorous quotations*. Oxford: Oxford University Press. pp. 229, 295.

7 Jones, T. and Palin, M. (1983). 'Every Sperm is Sacred', from the film *Monty Python's The Meaning of Life*. https://youtu.be/fUspLVStPbk.

8 Jarski, R. (2004). *The funniest thing you never said: The ultimate collection of humorous quotations*. London: Ebury Press. p.419.

9 Loyau, A., Petrie, M., Saint Jalme, M. & Sorci, G. (2008). Do peahens not prefer peacocks with more elaborate trains?. *Animal Behaviour*, 76, E5–E9.

10 Takahashi, M. *et al.* (2008). Peahens do not prefer peacocks with more elaborate trains. *Animal Behaviour*, 75, 1209–1219; Dakin, R. & Montgomerie, R. (2011). Peahens prefer peacocks displaying more eyespots, but rarely. *Animal Behaviour*, 82, 21–28; Loyau, A., *et al.* (2008). Do peahens not prefer peacocks with more elaborate trains?. *Animal Behaviour*, 76, E5–E9.

11 Buss, D.M. & Schmitt, D.P. (2019). Mate Preferences and their Behavioral Manifestations. *Annual Review of Psychology*, 70, 77–110.

12 Plomin, R. *et al.* (2016). Top 10 Replicated Findings from Behavioral Genetics. *Perspectives on Psychological Science*, 11, 3–23; Devlin, B., Daniels, M. & Roeder, K.

(1997). The heritability of IQ. *Nature*, 388, 468–471.

13 Feldman, M.W. & Ramachandran, S. (2018). Missing compared to what? Revisiting heritability, genes and culture. *Philosophical Transactions of the Royal Society B-Biological Sciences*, 373. 20170064.

14 Hills, T. & Hertwig, R. (2011). Why Aren't We Smarter Already: evolutionary trade-offs and cognitive enhancements. *Current Directions in Psychological Science*, 20, 373–377.

15 Ruch, W. (2008). Psychology of Humor. In: V. Raskin, ed. *The Primer of Humor Research*. Berlin & Boston, MA: De Gruyter, Inc.; Vernon, P. A. *et al.* (2008). Genetic and environmental contributions to humor styles: a replication study. *Twin Research and Human Genetics*, 11, 44–7; Baughman, H. M. *et al.* (2012). A Behavioral Genetic Study of Humor Styles in an Australian Sample. *Twin Research and Human Genetics*, 15, 663–667.

16 Greengross, G. & Miller, G. (2011). Humor ability reveals intelligence, predicts mating success, and is higher in males. *Intelligence*, 39, 188–192; Christensen, A. P. *et al.* (2018). Clever People: Intelligence and Humor Production Ability. *Psychology of Aesthetics Creativity and the Arts*, 12, 136–143; Jonason, P. K. *et al.* (2019). Is smart sexy? Examining the role of relative intelligence in mate preferences. *Personality and Individual Differences*, 139, 53–59.

17 Henrich, J., Heine, S. J. & Norenzayan, A. (2010). The weirdest people in the world?. *Behavioral and Brain Sciences*, 33, 61–83.

18 Greengross, G. & Miller, G. (2011). Humor ability reveals intelligence, predicts mating success, and is higher in males. *Intelligence*, 39, 188–192.

19 Gueguen, N. (2010). Men's sense of humor and women's responses to courtship solicitations: An experimental field study. *Psychological Reports*, 107, 145–156.

20 Wilbur, C.J. & Campbell, L. (2011). Humor in Romantic Contexts: Do Men Participate and Women Evaluate?. *Personality and Social Psychology Bulletin*, 37, 918–929.
21 Woolf, V. (1945). *A Room of One's Own*. Harmondsworth: Penguin.
22 Lippa, R. (2007). The Preferred Traits of Mates in a Cross-National Study of Heterosexual and Homosexual Men and Women: an examination of biological and cultural influences. *Archives of Sexual Behavior*, 36, 193–208. 10.1007/s10508–006–9151–2.
23 Hitchens, C. (2007). Why Women Aren't Funny. *Vanity Fair*, Vol 49, p.54; Tosun, S., Faghihi, N. & Vaid, J. (2018). Is an Ideal Sense of Humor Gendered?: a cross-national study. *Frontiers in Psychology*, 9, 199. 10.3389/fpsyg.2018.00199.
24 Robinson, D.T. & Smith-Lovin, L. (2001). Getting a laugh: gender, status, and humor in task discussions. *Social Forces*, 80, 123–158. 10.1353/sof.2001.0085.
25 Stewart-Williams, S. & Thomas, A.G. (2013). The Ape That Thought It Was a Peacock: does evolutionary psychology exaggerate human sex differences? *Psychological Inquiry*, 24, 137–168.
26 Williams, M. & Emich, K.J. (2014). The Experience of Failed Humor: implications for interpersonal affect regulation. *Journal of Business and Psychology*, 29, 651–668.
27 Bryant, G.A., *et. al.* (2018). The Perception of Spontaneous and Volitional Laughter Across 21 Societies. *Psychological Science*, 29, 1515–1525.
28 Vettin, J. & Todt, D. (2005). Human laughter, social play, and play vocalizations of non-human primates: an evolutionary approach. *Behaviour*, 142, 217–240.
29 Greengross, G. & Martin, R.A. (2018). Health among humorists: susceptibility to contagious diseases among

improvisational artists. *Humor-International Journal of Humor Research*, 31, 491–505.

30 Rotton, J. (1992). Trait Humor and Longevity: Do Comics Have the Last Laugh?. *Health Psychology*, 11, 262–266; Stewart, S. *et al.* (2016). Is the last 'man' standing in comedy the least funny?: a retrospective cohort study of elite stand-up comedians versus other entertainers. *International Journal of Cardiology*, 220, 789–793.

31 Papousek, I. (2018). Humor and Well-being: a little less is quite enough. *Humor-International Journal of Humor Research*, 31, 311–327.

32 Schneider, M., Voracek, M. & Tran, U.S. (2018). 'A joke a day keeps the doctor away?' Meta-analytical evidence of differential associations of habitual humor styles with mental health. *Scandinavian Journal of Psychology*, 59, 289–300.

33 Dunbar, R.I.M. *et al.* (2012). Social laughter is correlated with an elevated pain threshold. *Proceedings of the Royal Society B-Biological Sciences*, 279, 1161–1167.

34 Papousek, Humor and Well-being, op. cit.

Chapter Seven: Jokes and Culture

1 Wiseman, R. (2002). Laughlab: the scientific search for the world's funniest joke https://richardwiseman.files.wordpress.com/2011/09/ll-final-report.pdf. [Accessed 30 June 2019].

2 Wiseman, R. (2008). *Quirkology: The curious science of everyday lives*. London: Pan Books.

3 Wood, A. & Niedenthal, P. (2018). Developing a social functional account of laughter. *Social and Personality Psychology Compass*, 12. 10.1111/spc3.12383.

4 Sign in The Saracen's Head, Glasgow.

5 Fraley, B. & Aron, A. (2004). The effect of a shared

humorous experience on closeness in initial encounters. *Personal Relationships,* 11, 61–78.

6 Kashdan, T.B., Yarbro, J., McKnight, P.E. & Nezlek, J.B. (2014). Laughter with someone else leads to future social rewards: temporal change using experience sampling methodology. *Personality and Individual Differences*, 58, 15–19.

7 Fowler, J.H. & Christakis, N.A. (2008). Dynamic spread of happiness in a large social network: longitudinal analysis over 20 years in the Framingham Heart Study. *British Medical Journal*, 337. 10.1136/bmj.a2338.

8 Shalizi, C.R. & Thomas, A.C. (2011). Homophily and Contagion Are Generically Confounded in Observational Social Network Studies. *Sociological Methods & Research*, 40, 211–239.

9 Kramer, A.D.I., Guillory, J.E. & Hancock, J.T. (2014). Experimental evidence of massive-scale emotional contagion through social networks. *Proceedings of the National Academy of Sciences of the United States of America*, 111, 8788–8790.

10 Kirsch, A. (2007). A Poet's Warning. *Harvard Magazine*. Nov–Dec 2007. https://harvardmagazine.com/2007/11/a-poets-warning.html [Accessed 25 May 2019]. Auden reads the whole poem at: https://www.youtube.com/watch?v=JZE_bhSUgG8.

11 Coviello, L. *et al.* (2014). Detecting Emotional Contagion in Massive Social Networks. *Plos One*, 9. 10.1371/journal.pone.0090315; Baylis, P. *et al.* (2018). Weather impacts expressed sentiment. *Plos One*, 13. 10.1371/journal.pone.0195750.

12 Kross, E. *et al.* (2019). Does Counting Emotion Words on Online Social Networks Provide a Window into People's Subjective Experience of Emotion?: a case study on Facebook. *Emotion*, 19, 97–107. 10.1037/emo0000416.

13 Maurin, D., Pacault, C. & Gales, B. (2014). The jokes are vectors of stereotypes: example [sic] of the medical profession from 220 jokes. *Presse Medicale*, 43, E385–E392.

14 Quoted in Perez, R. (2016). Racist humor: then and now. *Sociology Compass*, 10, 928–938. 10.1111/soc4.12411.

15 Martin, R.A. & Ford, T.E. (2018). The Social Psychology of Humor. In: R.A. Martin & T.E. Ford, eds, *The Psychology of Humor: an integrative approach*. (Second Edition). London: Academic Press.

16 Thomae, M. & Viki, G.T. (2013). Why did the woman cross the road? The effect of sexist humor on men's rape proclivity. *Journal of Social, Evolutionary, and Cultural Psychology*, 7, 250–269.

17 Saucier, D.A. *et al.* (2018). 'What do you call a Black guy who flies a plane?': The effects and understanding of disparagement and confrontational racial humor. *Humour*, 31, 105–128.

18 Moalla, A. (2015). Incongruity in the generation and perception of humor on Facebook in the aftermath of the Tunisian revolution. *Journal of Pragmatics*, 75, 44–52.

19 Orwell (1968). Funny But Not Vulgar. In: *The Collected Essays, Journalism and Letters of George Orwell*. New York: Harcourt Brace Jovanovich. Originally published in *The Leader*, 28 July 1945. Cited by Moalla (2015), ibid.

20 Macnab, G. (2011). Rudolph Herzog: punchlines from the abyss. *Guardian*. 25 May 2011. https://www.theguardian.com/books/2011/may/25/rudolph-herzog-dead-funny. [Accessed 30 June 2019].

21 Lewis, B. (2008). *Hammer & Tickle: a history of communism told through communist jokes*. London: Weidenfeld & Nicolson.

22 Ibid.

23 Draitser, E. (1978). *Forbidden laughter: Soviet underground*

jokes. Los Angeles, CA: Almanac Publishing House.

24 Freud, S. (1905). *Jokes and their relation to the unconscious.* London: Hogarth Press and the Institute of Psycho-analysis.

25 Davies, C. (2002). *The mirth of nations*. New Brunswick, NJ: Transaction Publishers.

26 Cohen, T. (1999). *Jokes: Philosophical thoughts on joking matters*. Chicago, IL: University of Chicago Press.

27 Telushkin, J. (2002). *Jewish humor: What the best Jewish jokes say about the Jews*. New York: HarperCollins.

28 Ibid.

29 Baum, D. (2018). *The Jewish joke: An essay with examples (less essay, more examples).* London: Profile Books.

30 Sutton-Spence, R. & Napoli, D.J. (2012). Deaf jokes and sign language humor. *Humor*, 25. 10.1515/humor-2012–0016.

31 Ibid.

32 Chan, Y.C. *et al.* (2018). Appreciation of different styles of humor: An fMRI study. *Scientific Reports*, 8. 10.1038/s41598–018–33715–1.

33 Chen, H.-C. *et al.* (2013). Laughing at others and being laughed at in Taiwan and Switzerland. A cross-cultural perspective. In: J.M. Davis & J. Chey, eds. *Humour in Chinese Life and Culture: resistance and control in modern times.* Hong Kong: Hong Kong University Press. pp. 1–15.

34 Kobayashi, M. (2006). Senyrū: Japan's short comic poetry. In: J.M. Davis, ed. *Understanding humor in Japan.* Detroit, MI: Wayne State University Press. pp. 153–177.

35 Choy, H.Y.F. (2018). Laughable Leaders: a study of political jokes in mainland china. In: K.-F. Tam & S.R. Wesoky, eds. *Not Just a Laughing Matter: interdisciplinary approaches to political humor in China*. Singapore: Springer Singapore. pp. 97–115.

36 Haas, B. (2018). China bans Winnie the Pooh film after comparisons to President Xi. *Guardian*, 7 August 2018. http://www.theguardian.com/world/2018/aug/07/china-bans-winnie-the-pooh-film-to-stop-comparisons-to-president-xi. [Accessed 5 August 2019].

37 Quoted by: Hurley, M.M., Dennett, D.C. & Adams, R.B. (2011). *Inside jokes: using humor to reverse-engineer the mind.* Cambridge, MA: MIT Press.

Index